Springer Tracts in Modern Physics
Volume 143

Springer-Verlag Berlin Heidelberg GmbH

Springer Tracts in Modern Physics

Springer Tracts in Modern Physics provides comprehensive and critical reviews of topics of current interest in physics. The following fields are emphasized: Elementary Particle Physics, Solid-state Physics, Complex Systems, and Fundamental Astrophysics.
Suitable reviews of other fields can also be accepted. The editors encourage prospective authors to correspond with them in advance of submitting an article. For reviews of topics belonging to the above mentioned fields, they should address the responsible editor, otherwise the managing editor.

Elmar Schreiber

Femtosecond Real-Time Spectroscopy of Small Molecules and Clusters

With 131 Figures

Springer

Dr. habil. Elmar Schreiber
Max-Born-Institut
Rudower Chaussee 6
D-12474 Berlin
Email: eschreib@mbi-berlin.de
http://www.mbi-berlin.de

Physics and Astronomy Classification Scheme (PACS):
32.80.QK, 32.80.Rm, 33.80, 36.40

ISSN 0081-3869
ISBN 978-3-662-14759-7 ISBN 978-3-540-69663-6 (eBook)
DOI 10.1007/978-3-540-69663-6

Library of Congress Cataloging-in-Publication Data.
Schreiber, Elmar, 1957– Femtosecond real-time spectroscopy of small molecules and clusters / Elmar Schreiber.
p. cm. – (Springer tracts in modern physics; vol. 143) Includes bibliographical references and index.
1. Laser spectroscopy. 2. Laser pulses, Ultrafast. 3. Molecular dynamics. 4. Chemical kinetics.
5. Microclusters. I. Title. II. Series: Springer tracts in modern physics; 143. QC1.S797
vol. 143 [QC454.L3] 538'.6–dc21 97-49030

Originally published by Springer-Verlag Berlin Heidelberg New York in 1998.
Softcover reprint of the hardcover 1st edition 1998

Typesetting: Camera-ready copy by the author using a Springer TEX macro-package
Cover design: *design & production* GmbH, Heidelberg
SPIN: 10645161 56/3144-5 4 3 2 1 0 – Printed on acid-free paper

for Barbara

Preface

The progress of both technology and applications in the field of ultrafast processes within the last 20 years has been of remarkable dimensions. Not least because of the advent of all-solid-state femtosecond laser sources and because the extension of laser wavelengths by frequency conversion techniques has provided a variety of high-performance sources for extremely short light pulses. These excellent sources have enabled researchers all over the world extensively and quite successfully to investigate ultrafast phenomena in physical, chemical, and biological systems.

Femtosecond Real-Time Spectroscopy of Small Molecules and Clusters attempts to give a detailed overview of a small part of this new and exciting field situated at the boundary between physics and chemistry. The main subject of this book is research into the ultrafast dynamics of gas-phase molecules and clusters after excitation with intense femtosecond or picosecond laser pulses. Many textbook-like examples are presented.

This review was written both for graduate students, entering the fascinating world of femtosecond physics and chemistry and for researchers who want to get a vivid overview of state-of-the-art 'ultrafast' experimental techniques. The following topics are of central interest.

- Wave packet propagation phenomena influenced by electronic perturbations and by intramolecular vibrational redistributions (IVR); these phenomena are studied for extremely cold alkali dimers (K_2, Na_2) and trimers (K_3, Na_3). A highlight is the possibility to control actively the molecular dynamics of these species under certain experimental conditions, by means of ultrashort laser pulses.
- Ultrafast photodissociation dynamics, investigated for larger, very cold alkali aggregates $M_{n=3...10}$ (M = Na, K) as a function of excitation energy and cluster size. This allows detailed information on the stability of small alkali aggregates.
- The real-time observation of the structural relaxation of molecules after an electron has been detached by means of an ultrashort laser pulse, by applying a newly developed experimental method based on a charge reversal process. Structural relaxation times of optically excited clusters are determined. Here, the silver trimer acts as a model system.

To analyze the ultrafast dynamics of small molecules and clusters, the method of multiphoton ionization (MPI) spectroscopy combined with the pump&probe technique was chosen. Adiabatic expansion or sputtering was used to produce the molecules and clusters of interest. Quadrupole mass spectrometers enabled a mass-selective detection of the ionized species.

Investigations of molecular vibrational and photodissociation dynamics require high temporal resolution. Therefore, at the heart of the experimental setup are two ultrafast laser systems based on mode-locked titanium sapphire lasers with pulse widths of less than 80 fs operating at about 80 MHz repetition rate combined with a regenerative 1 kHz titanium sapphire amplifier. These laser systems allow the flexible and simultaneous operation of real-time experiments at both machines. The use of a synchronously pumped optical parametric oscillator combined with frequency-doubling greatly enlarges the spectral range of the femtosecond light sources. A brief description of both experimental setups, including the important experimental parameters, is given in Sect. 2.1.

The main topics listed above are discussed in the following chapters, starting with wave packet propagation phenomena (Chap. 3), followed by photodissociation dynamics (Chap. 4), and ultrafast changes of molecular configuration (Chap. 5). The results are summarized in Sect. 6.1. The forward view (Sect. 6.2) indicates some prospects for future investigations of the ultrafast spectroscopy of small molecules and clusters.

Finally, I would like to express my thanks to all those whose support, encouragement and criticism made this work possible.

- Professor Dr. Ludger Wöste for continuous support over the years and the opportunity to work in his group. All the experiments described here were performed in his laboratories. His great enthusiasm was an important motivation behind the results presented here.
- Professor Dr. Jörn Manz, from whom I learned the essentials of the theoretical description of the 'femtosecond world'. A fruitful cooperation with his research group enabled me to understand the fascinating details of the various wave packet propagation phenomena investigated.
- Professor Dr. Vlasta Bonačić-Koutecký, who gave me, during enlightening discussions, insight into the exciting theory of alkali clusters.
- Professor Dr. Steven R. Berry, whose extraordinary chemical knowledge made the 'NeNePo'[1] experiments so successful.
- Dr. Soeren Rutz, who joined me during his diploma and Ph.D. work. Soeren carried out many of the experiments presented in this book and it was his experimental talent and creativity that helped to make many of the results shown here possible.
- Prof. Dr. Regina de Vivie-Riedle and Dr. Birgit Reischl-Lenz, with whom I worked on the wave packet propagation phenomena of K_2 and Na_3, documented in several joint papers. It was a great pleasure.

[1] Abbreviation for a charge reversal process. For the curious reader: see Sect. 5.1.

- The Ph.D. students Dr. Katrin Kobe and Dr. Felix Holger Kühling, and also the diploma students Stefan Greschick, Jens Heufelder, Stefanie Rohland, Ansgar Ruff, Harald Ruppe, Georg Sommerer and David von Seggern. Each of them added an important piece to this work.
- Dr. Thomas Leisner and Dr. Sebastian Wolf for their cooperation on the 'NeNePo' experiment. In particular, they constructed the cluster source and cluster beam line for this experiment.
- Dr. Martin Garcia and Dr. Michael Hartmann, who gave me short courses on their theories of charge reversal processes, and Dr. Johann Gauss, who calculated the potential-energy surfaces of Na_3 and K_3.
- Dr. Pit Froben (FU Berlin), Dr. Claus-Peter Schulz, Dr. Georg Korn, and the Ph.D. students Christiana Bobbert and Parviz Farmanara from the Max-Born-Institut in Berlin for their critical reading and refining of the manuscript.
- All collaborators at the Institute for Experimental Physics of the Free University of Berlin, especially the electronics and mechanics workshop.
- Dr. Hans Kölsch, Jacqueline Lenz, Friedhilde Meyer, and Petra Treiber from Springer-Verlag for their always excellent cooperation.

I gratefully acknowledge continuous financial support from the Deutsche Forschungsgemeinschaft within project A8 of the Sonderforschungsbereich 337.

Berlin, February 1998 *Elmar Schreiber*

Contents

1. Introduction

The study of molecular dynamics in the femtosecond time domain (1 fs $= 10^{-15}$ s) can be regarded as the ultimate achievement in half a century of development of techniques for the exploration of the most elementary motion of atoms bound by chemical forces [1]. The outstanding books by Zewail, *Femtochemistry*, Vols. 1 and 2 [2], and Manz and Wöste, *Femtosecond Chemistry*, Vols. 1 and 2 [3], as well as the contributions [4, 5] presented at the two 'Femtochemistry' conferences in Berlin (1993) and Lausanne (1995), organized by Manz and Chergui respectively, document how femtosecond probing of molecular dynamics allows the viewing of new phenomena and the reaching of new frontiers. Applying real-time femtosecond spectroscopy to molecules or clusters allows us to make a 'movie' of their molecular dynamics revealing such phenomena as wave packet propagation, coherent control, internal vibrational redistribution, and ultrafast photodissociation or structural redistribution.

The central idea of this ultrafast spectroscopy of molecules and clusters is the preparation of molecular wave packets followed by the detection of their motion in real time. A wave packet is defined as a coherent superposition of several energy eigenstates [6–9]. To prepare a wave packet, laser pulses with a spectral width, ΔE_ℓ, that is broad compared to the level spacing, ΔE_{m}, of the molecular or cluster eigenenergies have to be applied. In the time domain, this implies that relative to the period $T = 2\pi\hbar/\Delta E_{\mathrm{m}}$ of the molecular motion considered, the duration of the preparation pulse has to be short. Vibrational periods of small molecules and clusters are of the order of 10^{-14} to 10^{-12} s [10]. With the availability of lasers generating pulses in the sub-100 fs regime [11–14], the real-time spectroscopy of these systems became feasible. By his introduction of femtosecond 'pump&probe' techniques [15–19], Zewail then established the direct observation of ultrafast molecular dynamics, which was named 'laser femtochemistry' [20].

Since then the ultrafast dynamics of a few molecular, especially dimer, systems have been studied. Zewail observed the vibrational and rotational revival of excited I_2 [21–23]. Stolow studied the same system, applying femtosecond pump&probe zero-kinetic-energy (ZEKE) photoelectron spectroscopy [24, 25]. Gerber observed fascinating features in the ultrafast dynamics of the sodium dimer's multiphoton ionization (MPI) (see Fig. 1.1 a), induced by high intensity femtosecond laser excitation [26–31]. The results are well described by theoretical simulations of Engel and Meier [32–36]. Recently, Girard presented a theory and an experiment on one-color coherent control for the closely related Cs_2 system [37–40]. In Li_2 Leone observed vibrational and rotational recurrences with single rovibronic control of an intermediate state [41].

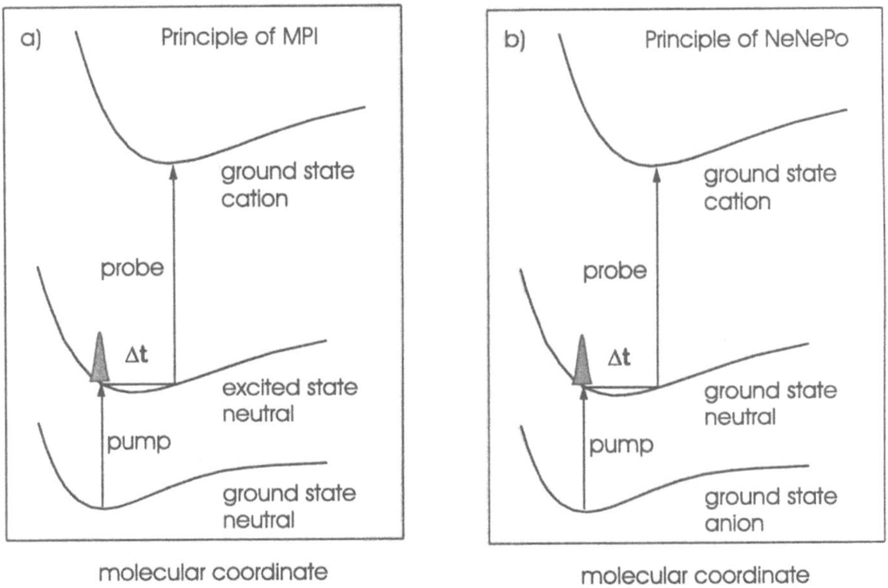

Fig. 1.1. Principles of the real-time multiphoton ionization (MPI) **(a)** and NeNePo **(b)** spectroscopic technique. **(a)** Principle of time-resolved MPI spectroscopy. A wave packet is prepared in an excited state of the neutral system by a pump pulse. Since in general the transition probability to the ion state is a function of the wave packet's location on the potential-energy surface, the propagation of the wave packet can be probed by a second, time-delayed pulse. **(b)** Principle of the time-resolved NeNePo process. Starting in the anion's potential-energy surface, an ultrashort pump pulse detaches an electron and prepares a wave packet in the neutral. After a certain delay time Δt a probe pulse photoionizes the neutral. The time-dependent signal of the cation's intensity is detected. For convenience, this method is called 'NeNePo', **N**egative-to-**N**eutral-to-**Po**sitive

In Sect. 3.1.1, for a similar system, the model molecule K_2 excited to its $A\,^1\Sigma_u^+$ state, the wave packet propagation is explored in greater detail by real-time three-photon ionization (3PI) spectroscopy. Applying laser pulses of 'moderate' intensities allows the selective detection of the pure vibrations of the A state, in excellent agreement with quantum dynamical calculations [42]. Both the favorable spectroscopic properties of K_2 and the special molecular dynamics induced by the selected moderate laser intensity combine to open a certain Franck–Condon window. The preferential transition pathway of the 3PI process can be determined easily.

It is often claimed that owing to the uncertainty principle, the gain in temporal resolution causes a loss in energy resolution. In Sects. 3.1.3 and 3.1.4 studies of two isotopes of K_2, as well as of the Na_2 molecule, each excited to their $A\,^1\Sigma_u^+$ state demonstrate that both high temporal and high spectral resolution can be obtained by femtosecond real-time spectroscopy. The coherence of the femtosecond pulse transferred to the dimer system enables the recovery of nearly all of the spectral information. Hence, Fourier analysis of the real-time spectra allows the identification in great detail of the excited vibrational levels of the A state. Energetic shifts due to spin–orbit coupling with a crossing 'dark' $b\,^3\Pi_u$ state show up. The achieved accuracy in time is close to 1 fs, and the spectral resolution is better than $0.1\,\mathrm{cm}^{-1}$. The high resolution in time reveals pronounced differences of the isotopes' dynamics (see also [42–50]).

To get even deeper insight into the laser-induced wave packet dynamics, visualization by spectrograms $I(\Delta t, \omega)$ [51–53] can be used. This technique nicely enables the direct observation of the time dependence (Δt) of the different frequency (ω) components originating from the propagating wave packet, including their relative contributions I. Revivals, total and fractional [54], are emphasized by spectrograms. Hence, as will be presented in Chap. 3 for dimers as well as trimers, the introduction of the spectrogram technique can be regarded as a promising method to analyze ultrafast molecular dynamics in greater detail.

The real-time spectra also reveal the dynamics of ultrafast intersystem crossing (ISC) processes [55–58] induced by spin–orbit coupling (see Sect. 3.1.3). Since continuous wave (cw) high-resolution spectroscopy inherently operates at the radiative equilibrium, its Fourier transformation is scarcely able, in practice, to deduce those elementary processes prepared by a Fourier-limited pulse. This becomes even more true for larger molecules as seen, for example, in the case of Na_3 excited to its electronic B state (see Sect. 3.2.2). The strong increase of a larger molecule's density of states opens up a great variety of intramolecular vibrational redistribution (IVR)[1] processes. Therefore, real-time spectroscopy, with its direct access to these processes, is able to find this information directly, whereas it is often difficult to obtain this knowledge by cw spectroscopy. The complexity of a trimer's vibrational modes can be considered as a prototype for the investigation of IVR (see Sect. 3.2.4).

Ever since the development of the laser, the dream of using light to control the future of matter has been of extraordinary interest to physicists and chemists. With the femtosecond laser technique this dream has come

[1] IVR is a description of how much initially localized energy in a specific mode of vibrational motion is shared among other types of accessible nuclear motion. Fundamentally, IVR is a manifestation of the loss of quantum coherence imposed on the initial state by the preparation process.

true.[2] Today the close interplay between theory and sophisticated laser experiments can be used specifically to manipulate the quantum behavior of atoms, molecules, and clusters.

The theory of quantum control pioneered by Tannor, Rice, and Kosloff [60, 61] allows, in principle, the computing of the optimal light field required to reach any specified quantum state of the sample under study. However, in practice there is no guarantee that one knows enough about the studied species, or has enough time to perform the computation [62], or even is able to generate the calculated optimal light pulse in the laboratory [63–65]. The current experimental approach to the control of molecular dynamics through the use of wave packets prepared by shaped laser pulses demands a detailed understanding of the various orders of molecular wave packet phenomena, where the spectrogram technique (Chap. 3) might be a most promising tool.

The first examples of experiments of this type [19, 29, 30, 37, 66, 67] have proved the ideas of Tannor, Rice, and Kosloff [60, 61]. Most of these pump&control experiments were carried out on diatomic molecules. In larger systems with three or more vibrational degrees of freedom, the situation becomes much more complicated and it is a fascinating question whether the concept of 'controlled molecular dynamics' can still be realized (see Sect. 3.2.4).

Two examples which emerged from a fruitful cooperation of theory and experiment are discussed in this book. First, once again the K_2 molecule excited to its A state acts as a model system (Sect. 3.1.5). Its special spectroscopic properties combined with the dynamics induced by 'femtosecond state preparation' facilitate the transition from pump&probe to control spectroscopy. The intensity of the pump pulse serves as the control parameter, allowing the forcing of the molecule to perform either the A state vibrations or its ground state dynamics.

The general complementarity of sensitivities in cw and femtosecond spectroscopy has been anticipated by Zewail [66] and it is verified for the Na_3 molecule excited to its electronic B state (see Sect. 3.2.4). This system has already been studied in great detail by various experimental and theoretical techniques such as cw two-photon ionization spectroscopy [68–70], femtosecond pump&probe spectroscopy at high intensities [29, 30, 71], quantum ab initio studies [72–74], two-dimensional simulations of the pseudorotational progressions in the cw absorption spectra [75–78], and, finally, three-dimensional simulations by means of empirical potential-energy surfaces (PESs) [79, 80].

Questions concerning IVR processes in the trimer, however, were by no means answered satisfactorily. As presented in this book (Sect. 3.2.4), the effect of selective state preparation can be achieved by variation of the exciting pump pulse duration. The possibility of preparing specific vibrational modes by ultrafast pulses provides a detailed understanding on how the in-

[2] Another promising approach to controlling processes such as chemical reactions is presented by Brumer and Shapiro [59], using the coherence of lasers by applying nanosecond lasers with close to transform-limited bandwidth.

duced molecular oscillation is built up. Starting in the dominantly prepared symmetric stretch mode, the sodium trimer finds its pseudorotational rhythm after a few picoseconds owing to an ultrafast IVR process. It has to be emphasized that neither in picosecond nor in cw spectroscopy is the trimer's symmetric stretch mode observable. These complementary experimental results for cw/picosecond and femtosecond excitation are fully verified and explained by the first consistent three-dimensional quantum dynamical ab initio study of pump&probe experiments (see [62, 81]). This leads to the second model system where active control, here pulse-width-controlled molecular dynamics of Na_3, can be demonstrated (Sect. 3.2.4).

Concerning the observation of wave packet propagation phenomena, the K_3 molecule is another promising candidate. Theoretical calculations [82] predicted an electronic state comparable to the Na_3 B state at about 800 nm. Different, highly sensitive methods, such as MPI and depletion spectroscopy, were applied but failed [83]. However, as presented in Sect. 3.2.5, with femtosecond real-time spectroscopy it is possible to observe both the vibrational and the dissociation dynamics of this system. This, once again, is an example of the complementarity of cw and femtosecond spectroscopy.

The K_3 system represents a fascinating limiting case, where both wave packet propagation phenomena and dissociation dynamics of the molecule can be analyzed. This leads to a new set of investigations.

Photodissociation, especially of small molecules and clusters, can be regarded as the motor for many important chain reactions. To determine the characteristics of this 'motor', the exploration of the real-time dissociation of small elemental clusters induced by ultrashort light pulses is of great help. It will give a fundamental insight into the stability and chemical forces of these species.

Compared with solids, elemental clusters are characterized especially by a large number of surface atoms. Therefore, generally, clusters are different from both molecules and solids, often presenting very specific properties [84–92] and size effects [93]. Owing to the inherent instability of clusters, general information about their fundamental properties has become available only recently, using new experimental techniques. Here, catalysis [94–98], solvation [99, 100], reactivity [15, 17, 18, 101, 102], energy transfer [93,103–105], and fragmentation [106–111] processes are some of the main areas of interest.

Small alkali clusters play an outstanding role in fundamental research on clusters since they can be studied rather easily theoretically (e.g. [89, 112, 113]) and produced experimentally [114–116]. The complexity of these clusters is sufficient to obtain manifold new insights into the transition from molecules to bulk [117].

One essential lack of empirical knowledge is of the stability of these clusters themselves [118–121]. To investigate their stability, real-time studies of the fragmentation probability are required. The lifetime of excited clusters is expected to be in the range of 10^{-10} to 10^{-14} s [10, 122, 123]. Therefore, no matter how sophisticated experiments with conventional nanosecond-spectroscopy [124–127] are, they are inherently unable to resolve the real-time dynamics of the dissociation process. As a consequence, direct probing of the temporal evolution of the dissociating system requires a pump&probe technique in the pico- or femtosecond time domain. The pump pulse now prepares a wave packet in a repulsive state of the cluster. The probing can be accomplished by exciting the transient cluster either to an upper electronic state or into the ionization continuum. The temporal evolution of the laser-induced fluorescence (LIF), absorption, or ion signal then directly reflects the real-time dependence of the evolving wave packet. Fundamental examples of the real-time laser-induced fluorescence of repulsive systems, neatly explaining the concept and principal methodology, are given in [17, 128–130].

To observe ultrafast fragmentation of excited alkali clusters the appropriate tool is real-time MPI spectroscopy. This technique allows the mass-selected detection of the ultrafast photodissociation with high sensitivity. In 1992 Gerber and coworkers presented the first femtosecond time-resolved experiments in cluster physics [32, 131, 132], showing differences in the fragmentation behavior of $Na_{n\leq 21}$ clusters dependent on the excitation at different wavelengths.

In this book the real-time photodissociation dynamics of small sodium ($Na_{n=3...10}$) and potassium ($K_{n=3...9}$) clusters are studied as a function of cluster size as well as excitation wavelength (Sects. 4.2–4.4). The ratio of dissociative to radiative decay is a measure of the predissociation of an electronic state [122, 133]. For the C state of Na_3 the electronic predissociation dynamics and especially the localization of its onset are analyzed in detail by ultrafast spectroscopy (Sect. 4.1).

A central application of ultrafast dynamics pertains to configurational changes and IVR in vertically excited or ionized electronic-vibrational states of small molecules and clusters. However, the ultrafast dissociation often prevents information concerning the vibrational dynamics of the studied system from being obtained. Here a new approach called 'NeNePo' (see Fig. 1.1 b) opens a window to acquire a deeper insight into the dynamics of a cluster. This technique, based on a charge reversal process, enables the direct preparation of a wave packet in the ground state of the cluster. An ultrashort pump pulse interacts with an anion to produce by photodetachment a neutral cluster excited to several vibrational levels of its ground state. Unlike to the preparation of a wave packet via a resonant impulsive stimulated Raman scattering process (RISRS) (see Sect. 3.1.5), a selective excitation in the ground state becomes possible here. Moreover, the wave packet propagation is not superimposed on any dynamics of intermediate electronic states. Hence, the neutral's pure ground state dynamics can be probed by a subsequent laser pulse.

The first results are presented in Sect. 5.1, revealing information about ultrafast structural relaxation times of the prepared molecule or cluster. The results are in excellent agreement with an elegant theoretical description given by Bennemann and coworkers [134, 135]. They combine an electronic theory with molecular dynamics calculations to study the ultrafast structural response of optically excited small clusters. From this the time dependence of the ionization potential of small clusters upon photodetachment can be estimated. Jortner, Bonačić-Koutecký, and coworkers [136] use another approach, applying the density matrix method to investigate laser-pulse-induced charge reversal processes in small silver molecules and clusters. Principal access to the real-time observation of isomerization processes, for example, is possible by this new experimental technique.

To observe all these exciting phenomena in real time it is essential to use a suitable state-of-the-art ultrafast laser system (see Sect. 2.1). Therefore, I should like to finish this chapter with a few comments on ultrafast laser systems.

During the past three decades, we have seen dramatic progress in the generation of ultrafast laser pulses and their application to the study of phenomena in many disciplines [10, 13, 137–150]. As a consequence a kind of 'femtofascination' [14] has arisen and forms the basis of the so-called 'femtosecond spectroscopy'.

More recently, the development of femtosecond real-time spectroscopy has been greatly improved by the availability of the titanium sapphire laser technology [148–150]. The discovery of self-mode-locking or Kerr-lens mode-locking [151, 152] has opened the way to benefitting efficiently from the titanium sapphire's enormous optical bandwidth ($\sim$ 100 nm [153]) for ultrashort pulse generation. This, combined with the appearance of novel techniques for dispersion compensation [154–158], enables the generation of nearly transform-limited pulses in either the picosecond or the femtosecond regime [159–162]. Owing to its excellent mechanical and thermal properties, the pulse-to-pulse fluctuations are extremely low, promoting this laser as an ideal tool for research in the fascinating field of ultrafast molecular processes induced by light.

Compared to previously used colliding-pulse mode-locked (CPM) dye lasers [163], mostly combined with excimer, pumped dye amplification dye cells [164–166] and often followed by white-light continuum generation [167], the regeneratively mode-locked titanium sapphire laser [159] is relatively easy to handle. Its broad tuning range ($\sim$ 700 nm to 1.1 μm), in combination with different nonlinear optical techniques such as optical harmonic generation [168–170] or optical parametric oscillation [171, 172], covers the whole spectral region from the far infrared ($\sim$ 10 μm) [173] to the ultraviolet ($\sim$ 150 nm) and is a comfortable basis for the spectroscopy of molecules and clusters. Beyond this, the titanium sapphire medium is also well suited to extracavity amplification [174–177]. Hence, highly nonlinear processes such as MPI spectroscopy, as well as molecular and cluster systems of very low density, can be studied with sufficient sensitivity.

These characteristics illustrate why two titanium sapphire lasers with an optional regenerative amplifier served as the basis of nearly all the investigations presented in this book.

2. Femtosecond Real-Time Spectroscopy

Femtosecond real-time spectroscopy requires sophisticated experimental and theoretical techniques. Some experimental basics are given in this chapter. A few remarks on the theoretical basics relevant to the investigations presented in this book are included in the second part of this chapter. It should be pointed out that this treatment is intended to give a rapid overview but is not at all complete. Those readers who are looking for a more detailed description of these fascinating experimental and theoretical techniques should therefore consult the references given in Sects. 2.1 and 2.2.

2.1 Experimental Basics

The principles of two experimental setups are described in this section. In Sect. 2.1.1 the setup for the real-time MPI experiments is introduced. With this setup, the investigations discussed in Chaps. 3 and 4 were carried out. The experimental basics are concluded with a brief description (Sect. 2.1.2) of the setup used for the real-time studies of charge reversal processes discussed later in Chap. 5.

2.1.1 Setup for the Real-Time MPI Experiments

Resonantly enhanced multiphoton ionization (REMPI), more generally called multiphoton ionization (MPI), of small molecules and clusters has been studied in recent years by a variety of techniques and is rather well understood. Different alkali molecules and clusters have been investigated by applying real-time MPI spectroscopy. The time evolution of the MPI signals is obtained by means of either picosecond or femtosecond pump&probe techniques followed by mass-selective detection of the ionized aggregates.

The principle of real-time MPI spectroscopy is summarized schematically in Fig. 2.1. The figure shows the two-photon ionization (TPI) of a molecule for the simplest case. Three potential-energy surfaces (PESs) of the molecule, the ground and excited electronic states of the neutral molecule and the ground state of the cation, are shown. A bound excited electronic state is populated by the first (pump) pulse. As a result of the rather broad spectral width of the

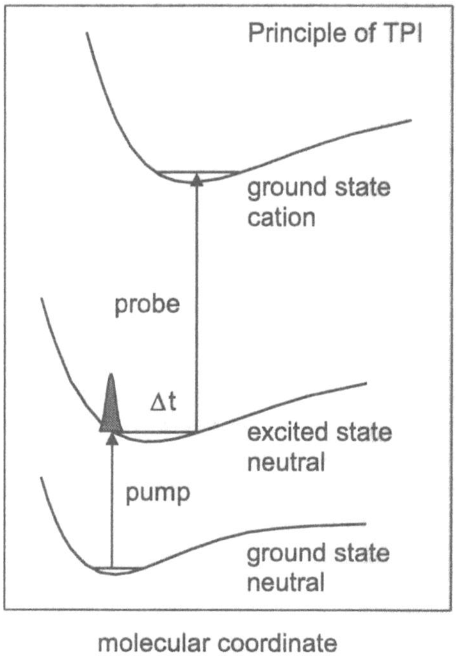

Fig. 2.1. Schematic illustration of the transient two-photon ionization (TPI) process of a molecule. Starting initially from the neutral molecules's ground state, a wave packet is generated in one of the molecule's excited electronic states by an ultrashort (pump) pulse. The prepared wave packet propagates on this PES, and after a certain delay time Δt a second laser (probe) pulse is applied to ionize the molecule

ultrashort laser pulse, the pump pulse simultaneously excites several vibrational levels of the excited state. Hence, a wave packet is generated that can propagate on the PES of the excited molecule. After a certain delay time Δt a second ultrashort pulse, the probe pulse interacts with the molecule, trying to ionize it. Varying gradually the delay time Δt between the pump and probe pulses allows the probing of the probability to ionize the so-prepared (excited) molecule as a function of time. Since in general the transition probability from the excited state to the ion state is a function of the wave packet's location on the PES, the temporal evolution, i.e. the propagation of the wave packet, can be directly probed by the second laser. Therefore, the temporal evolution of the ion intensity generated by the probe pulse can be regarded as a mapping of the wave packet's propagation. In other words, with real-time MPI spectroscopy it is possible – at least in some special cases – to make a movie of the ultrafast dynamics of molecules photoexcited by an ultrashort laser pulse. Some model examples are given in Chaps. 3 and 4.

To realize the real-time MPI spectroscopy of small alkali molecules and clusters an experimental setup as shown schematically in Fig. 2.2 is used. First, the main components of the setup, the laser systems used, the source for the molecules and clusters, and the mass-selective detection system are briefly described. An explanation of a few further details and the technique of the measuring process follows.

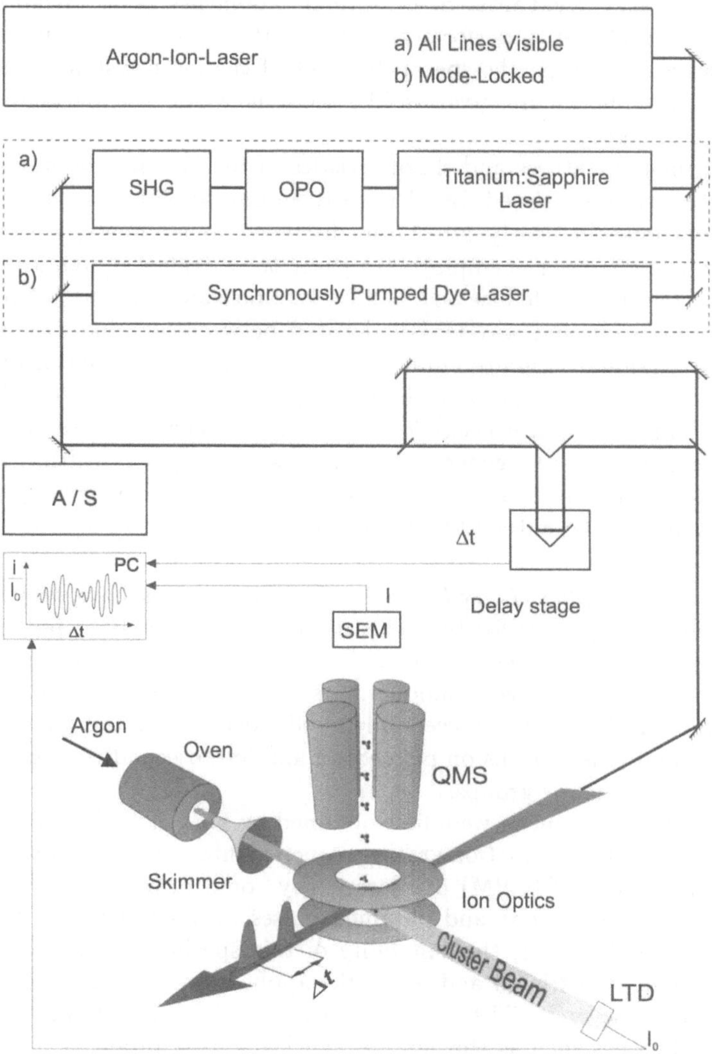

Fig. 2.2. Experimental setup (as used for the investigations on Na_3B). An argon ion laser (ps: mode-locked; fs: all lines, visible) pumps either a femtosecond laser system **(a)** (OPO: synchronously pumped optical parametric oscillator; SHG: second-harmonic generator) or a picosecond laser system **(b)** (taken from [178]). The pulse duration and spectral width of the laser pulses are measured by an autocorrelator (A) and a spectrometer (S) respectively. A Michelson arrangement allows the probe pulses to be delayed (Δt) with respect to the pump pulses. A quadrupole mass filter (QMS) enables the selection of the ensemble of investigated molecules ionized by a pump&probe cycle. A secondary electron multiplier (SEM) detects the intensity I of the ions as a function of the delay time Δt. A Langmuir–Taylor detector (LTD) measures the total intensity I_0 of the cluster beam. The ratio I/I_0 as a function of the delay time Δt is called the real-time spectrum

The Ultrafast Laser Systems. Since several modifications of ultrafast laser systems are employed to perform real-time MPI investigations on different molecules and clusters, the special features of the laser systems used in the various experiments are given in the sections where the particular experiments are discussed.

Here, the principal features and characteristics of the ultrafast laser systems used are briefly summarized. Besides the titanium sapphire laser which acts as the workhorse in nearly all of the discussed experiments, a synchronously pumped dye laser is employed to study the ultrafast dynamics of Na_3 on a picosecond timescale (see Sect. 3.2.2). For measurements with femtosecond time resolution and wavelengths located between 600 and 625 nm a synchronously titanium sapphire pumped optical parametric oscillator followed by frequency doubling is used. To investigate the Na_3 C state, two mode-locked titanium sapphire lasers have been synchronized. In all cases the essential parameter of the generated laser pulses, the pulse width, has to be determined. This problem is solved by an autocorrelation technique. Hence, the principles of an autocorrelator are briefly described at the end of this section.

Synchronously Pumped Mode-Locked Dye Laser. Organic dyes have proven to possess excellent properties for the generation of ultrashort laser pulses. Numerous approaches have been taken to the mode-locking of a dye laser, including both active and passive techniques.[1] As a result, tunable continuous wave (cw) dye lasers have been successfully mode-locked over the past 25 years [183, 184], and pulse widths on picosecond and femtosecond timescales have been reported by many groups.

The shortest dye laser pulses were first obtained by Fork et al. [12]. Placing a combined prism and diffraction grating sequence into the cavity of their colliding-pulse mode-locked (CPM) dye laser, they could provide quadratic and cubic phase compensation[2] and obtained pulses of 6 fs full width half maximum (FWHM). However, the tunability of this special dye laser configuration is extremely restricted and limits the application to spectroscopic investigations rather severely. Therefore, sometimes 'old-fashioned' and less sophisticated techniques such as the 'simple' synchronous pumping of a dye

[1] Basics on the subject of mode-locking of lasers are found in, for example, the well-known articles of Smith [179], Harris and McDuff [180], and Kuizenga and Siegman [181], as well as in the overview paper of New [182].

[2] To generate pulses much shorter than a picosecond, it is necessary to take account of group velocity dispersion (GVD) – the phenomenon which causes different frequency components to travel at different speeds, thereby broadening the pulses as they circulate in the laser cavity. For detailed overviews see [158, 185, 186]. GVD in an optical medium with a wavelength-dependent refractive index $n(\lambda)$ is determined by the second derivative $\mathrm{d}^2 n(\lambda)/\mathrm{d}\lambda^2$ of the dispersion curve $n(\lambda)$. The GVD causes temporal reshaping of wave packets (laser pulses). This can be a broadening or a shortening shape change, depending upon the initial conditions (chirp) of the wave packet spectrum. The term 'chirp' means that the frequency of a wave packet is changing with time as in the chirping of a bird.

laser with a mode-locked argon laser are the better choice to perform a particular experiment. Synchronous pumping is a well-recognized technique for short pulse generation. It is a form of amplitude-modulation mode-locking where the gain is the pulse-shaping mechanism. In other words, the technique relies on round trip cavity times coinciding with new pump pulse arrivals [182, 187–189]. Any cw laser that can be excited by a mode-locked cw laser can obviously be synchronously pumped. If the pulse energy of the slave laser is sufficient to saturate the gain, then the laser will become synchronously mode-locked. In practice this means that all laser dyes which can be pumped by cw mode-locked ion lasers or Nd:YAG lasers are available for cw mode-locking.

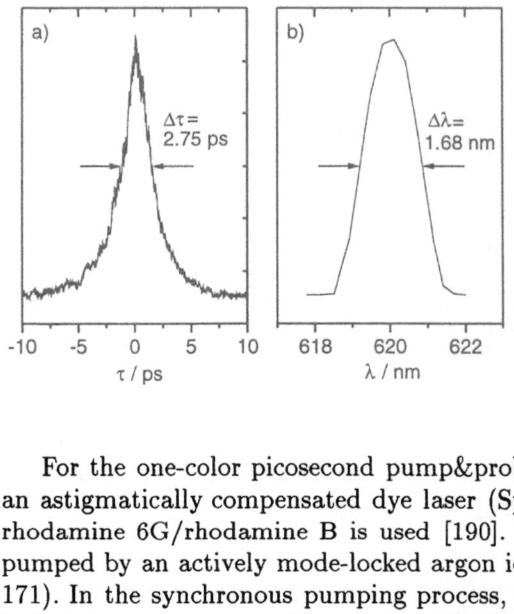

Fig. 2.3. Autocorrelation trace (**a**) and wavelength spectrum (**b**) of the synchronously pumped dye laser. Assuming single exponential decay , $\Delta t = 1.38\,\text{ps}$ (FWHM) is estimated. A bandwidth $|\Delta\bar{\nu}| = 40\,\text{cm}^{-1}$ is revealed from the spectrum

For the one-color picosecond pump&probe experiments (see Sect. 3.2.2), an astigmatically compensated dye laser (Spectra Physics model 375) with rhodamine 6G/rhodamine B is used [190]. The dye laser is synchronously pumped by an actively mode-locked argon ion laser (Spectra Physics model 171). In the synchronous pumping process, the output pulses of the mode-locked ion laser are used to excite the dye laser, whose cavity length has been extended to be equal to the ion laser's cavity ($\sim$ 1.8 m) [187]. The transmission of the dye laser's output coupler is $\sim$ 22%. With this setup the emitted dye laser pulses have a pulse duration[3] of 1.38 ps and a bandwidth of $40\,\text{cm}^{-1}$ (see Fig. 2.3). The pulse repetition rate is $\sim$ 82.5 MHz and an average power of 120 mW is obtained over the tuning from 600 nm to 630 nm.

Mode-Locked Titanium Sapphire Laser. In the late 1980s a new highly tunable laser material, Ti^{3+}-doped sapphire ($Ti{:}Al_2O_3$), set out to conquer the world of ultrafast pulses. It looks as if the titanium sapphire laser is set to replace ultrafast dye lasers in almost all applications. The extraordinary

[3] The pulse width $\Delta\tau$ is estimated by an autocorrelator. The principle of measuring ultrashort pulse widths with autocorrelators is explained later in this section.

spectroscopic and laser characteristics of Ti:Al_2O_3 ion were first reported by Moulton [191, 192]. The Ti^{3+} ion is responsible for the laser action of titanium sapphire. The electronic ground state of the Ti^{3+} is split into a pair of vibrationally broadened levels as shown in Fig. 2.4 a. Absorption transitions occur over a broad range of wavelength from 400 nm to 600 nm. Fluorescence transitions occur from the lower vibrational levels of the excited state (^2E_g) to the upper vibrational levels of the ground state ($^2T_{2g}$). The upper-state lifetime is in the region of 3 μs [193]. The fluorescence band extends from 600 nm to wavelengths greater than 1 μm. However, laser action is restricted to wavelengths above 660 nm, since the long-wavelength side of the absorption overlaps with the short wavelength end of the fluorescence spectrum[4] (Fig. 2.4 b). Nevertheless, compared with dye lasers the ultra-broadband vibronic titanium sapphire laser has about four times as broad a tuning range.[5]

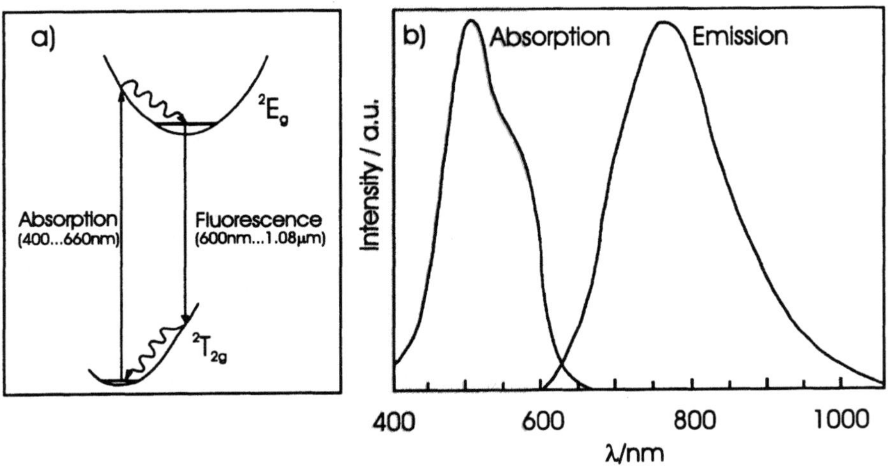

Fig. 2.4. (a) Energy level structure of Ti^{3+} in sapphire. (b) Absorption and emission spectra of titanium sapphire (by courtesy of Spectra Physics)

There are several other properties of titanium sapphire which differ greatly from dye lasers. Primarily, Ti:Al_2O_3 has a much smaller pump absorption cross section than organic dyes. Therefore, typical absorption lengths are a few millimeters up to several centimeters and the pump and laser beam normally have to be focused to a beam waist of ~ 50 μm diameter in the

[4] Throughout this book the abbreviation 'a.u.' is used for arbitary units.

[5] In the ideal case the length of mode-locked pulses generated by any laser, whether it is homogeneously or inhomogeneously broadened, can approach a value $\Delta\tau \sim 1/\Delta\nu$, where ν is the width of the gain profile. Consequently, gain media with large values of $\Delta\nu$ are desirable for the generation of the shortest mode-locked pulse.

crystal. The longer lengths of the gain media, in turn, increase the magnitude of nonlinear effects caused by the intense pulses passing through the crystal, as well as the amount of group velocity dispersion (GVD) within the cavity. These characteristics, combined with the large lifetime of the upper laser level (allowing high inversion), greatly affect the design of resonators for mode-locked titanium sapphire lasers and make new mode-locking techniques necessary (see e.g. references in [159]).

The mode-locking mechanism for the titanium sapphire laser can be described by the optical Kerr effect (OKE). Owing to the nonresonant bound-electron nonlinearity of the gain medium, the OKE can be exploited in the laser cavity to simulate a saturable absorber with a virtually instantaneous absorber recovery. However, the picture of 'saturable absorber' is not strictly correct since there is no saturation or depletion of a population across the transition. The OKE leads rather to a nonlinear intensity-dependent refractive index in the optical elements of the laser cavity, given by

$$n = n_0 + n_2 I. \tag{2.1}$$

Since the phase delay experienced by the propagating optical signal is proportional to the refractive index, the OKE causes a nonlinear intensity-dependent delay in addition to the linear contribution.[6] The intensity of a pulse varies across both its temporal and its spatial profile. Therefore, different parts of the profile will experience a different refractive index. As a result, the OKE provides a modulation function that follows the pulse profile and continues to strongly shape the pulses on each cavity round trip until the pulse is compressed to its final steady-state value. This mechanism leads to 'self-mode-locking' caused by the effects of self-phase-modulation (SPM)[7] [194, 195] and self-focusing.[8] Both phenomena are a consequence of different parts of an optical signal experiencing different refractive indices and thus different nonlinear phase shifts.

As SPM and GVD from the many dispersive elements within the laser cavity produce a positive chirp[9] on the pulse, some method must be employed

---

[6] For beam waists of $\sim 50\,\mu$m diameter, a power of $\sim 2.5\,$MW is required over a propagation length of 1 cm to produce of π non-linear shift at 800 nm.

[7] SPM occurs in the temporal domain. Considering a pulse, the more intense peak experiences a higher refractive index than the wings of the pulse. During propagation, this results in a changing phase across the pulse profile. The differential of phase with respect to time is a frequency change and so SPM results in pulses with frequency sweeps, so-called 'chirps'.

[8] Self-focusing occurs in the spatial domain and exploits the fact that the laser beam profile, usually a Gaussian TEM_{00} mode, experiences a changing refractive index profile across its diameter. This is analogous to propagation in a graded-index lens and thus results in focusing.

[9] A pulse is said to be positively chirped if its instantaneous frequency increases from leading edge to trailing edge. This is the type of chirp which will normally be imparted to a pulse after traversing 'normal' materials, i.e. with an upward curvature of the dispersion to the blue end. Similarly, a pulse is said to be

to enable the 'slow frequencies or wavelengths' to catch up with the faster ones. Otherwise, the cumulative effect of even a small chirp per round trip would create broadening and pulse substructure. Thus, an element is required which has negative GVD, i.e. an element which in principle can compensate the sum of the positive chirps mentioned above. The net GVD of prism combinations with a proper choice of prism material and distance between them [155], as well as of the Gires–Tournois interferometer (GTI) [196–199], can be negative. These elements, placed in the cavities of the titanium sapphire laser, enable the formation of the ultrashort pulses. For further information on the basic mode-locking process of titanium sapphire lasers see e.g. [186, 200–202].

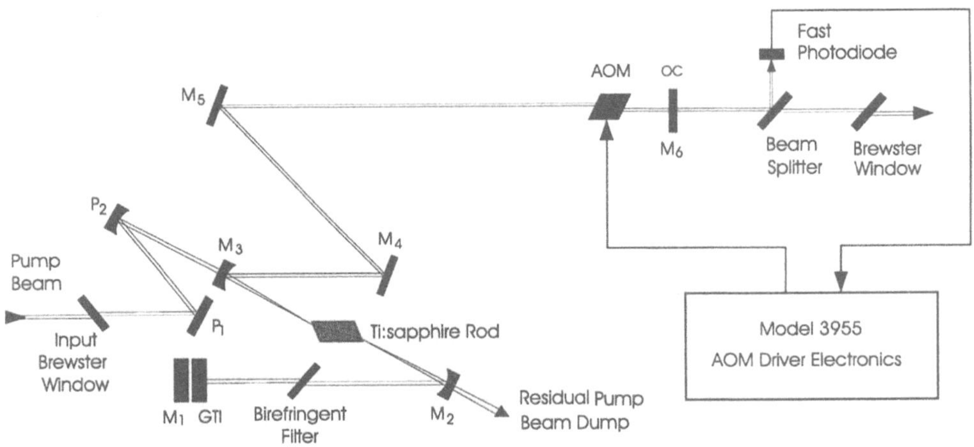

Fig. 2.5. Picosecond configuration of the folded titanium sapphire laser cavity. AOM acusto-optic modulator; OC output coupler; P_1, P_2 pump mirrors; M_1 to M_6 cavity mirrors; GTI Gires–Tournois interferometer (by courtesy of Spectra Physics)

The most significant difference between the different mode-locking techniques applied is the method employed to initiate the mode-locking [186]. The titanium sapphire lasers (Spectra Physics Tsunami model 3950B and 3960) used for the experiments described in this book are based on the regenerative mode-locking technique [159]. As known from active mode-locking techniques, a precise match is required between the resonance frequency of the optical cavity and the frequency of the radio-frequency (rf) signal driving an intracavity modulator. In a regeneratively mode-locked laser, the cavity acts as its own timing source. A photodiode senses the resonance frequency ($\sim$ 82 MHz) of the optical cavity and its output is conditioned to drive an acousto-optic modulator. Thus, the cavity resonance and rf driver frequen-

negatively chirped if its red spectral components have been retarded with respect to the blue ones.

cies are interdependent and the laser pulse performance operates optimally for extended periods of time without the dropouts or shut-downs associated with standard mode-locking systems.

A large frame argon ion laser with BeamLok^{TM} (Spectra Physics 2080, all line, visible), operating in the power mode, is used to pump the titanium sapphire laser usually, with $\sim$ 12 W output power. Two cavity configurations are used for the experiments, one for picosecond and the other for femtosecond operation [159]. The main difference is in the way of compensating for GVD to form nearly transform-limited pico- and femtosecond pulses. Fig. 2.5 shows the picosecond configuration. A GTI is employed [197–199]. Wavelength tuning is perfomed by a two-plate birefringent filter with a free spectral range of $\sim$ 150 nm. When configured for femtosecond operation prisms are used instead of the GTI and birefringent filter (Fig. 2.6). The prism configuration results in a collimated beam at a slit position. The movable slit can be positioned to select the desired wavelength. To cover the operating wavelength range (720 nm to 1080 nm) three partially overlapping mirror sets are used.

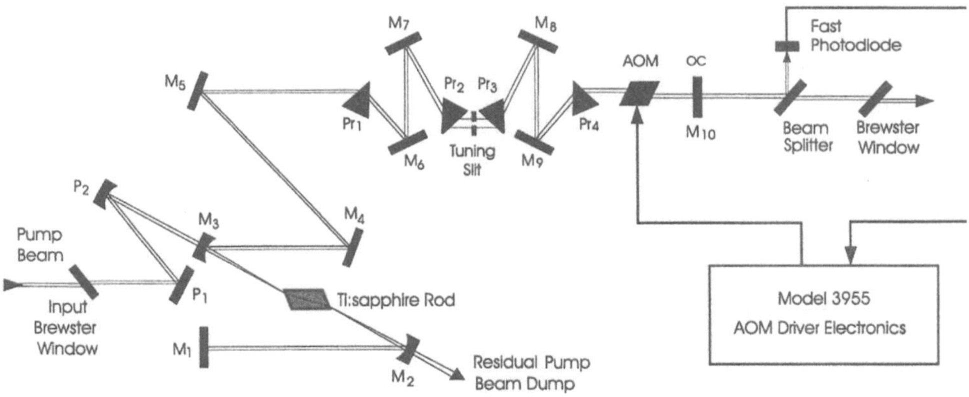

Fig. 2.6. Femtosecond configuration of the titanium sapphire laser cavity. AOM acusto-optic modulator; P_1, P_2 pump mirrors; M_1 to M_{10} cavity mirrors; PR_1 to PR_4 prisms, OC output coupler (by courtesy of Spectra Physics)

For the femtosecond configuration of the titanium sapphire laser a typical autocorrelation trace of the output pulses is shown in Fig. 2.7 a. Assuming a sech^2 shape, the duration of the emitted light pulses is about 70 fs (FHWM). With a bandwidth of about 11.6 nm (181 cm^{-1} FWHM) at a central wavelength close to $\lambda = 800$ nm, the pulses are approximately 1.2 times the Fourier limit (assuming a sech^2 shape). The pulse repetition rate is $\sim$ 80 MHz. Hence, every 12.5 ns a pulse is emitted. A tuning range from 700 nm to 1.1 μm can be obtained by means of three different mirror sets. The interferometric autocorrelation trace, given in Fig. 2.8, reveals even more detailed information (e.g. of a chirp) of the laser pulse train.

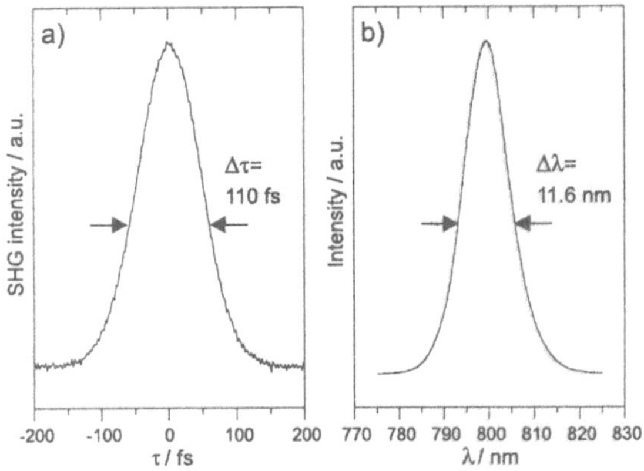

Fig. 2.7. Autocorrelation trace **(a)** and spectrum at $\lambda = 800\,\text{nm}$ **(b)** of the femtosecond titanium sapphire laser. Assuming a sech^2 pulse shape, a pulse width $\Delta t = 71\,\text{fs}$ (FWHM) is measured. The spectrum reveals a bandwidth $|\Delta\bar{\nu}| = (1/\lambda^2)\Delta\lambda = 181\,\text{cm}^{-1}$. With these data the pulse is 1.2 times bandwidth-limited (see Table 2.2)

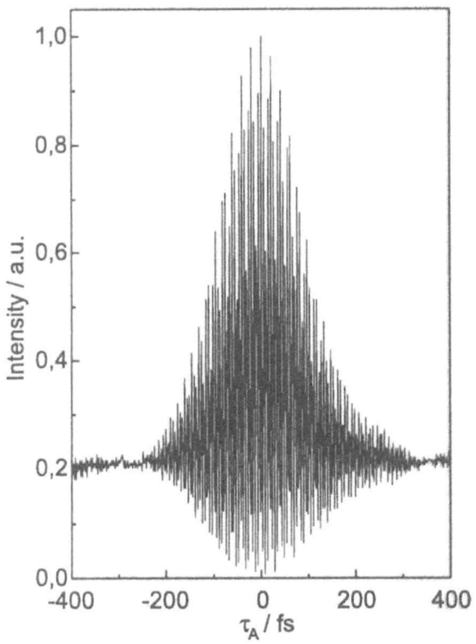

Fig. 2.8. Interferometric autocorrelation trace of the femtosecond titanium sapphire laser (taken from [203]). The pulse width is 70 fs. The trace was recorded with a step width of 1 fs

Synchronously Pumped Optical Parametric Oscillator. Optical parametric oscillators (OPOs) were first used in the mid-1960s as an alternative to dye lasers for generating coherent radiation tunable over a wide wavelength region [204, 205]. It is only recently, however, that OPOs have become a practical reality with the advent of new high-quality, nonlinear optical materials and high-power, mode-locked pump sources [206–211].

An OPO uses a nonlinear optical crystal in an oscillator cavity to generate new frequencies (called signal and idler) from a coherent input source (pump laser). The nonlinear optical response of the material can be described by the power series expansion of the optically induced polarization

$$P = \varepsilon_0(\chi^{(1)} \cdot E + \chi^{(2)} \cdot E \cdot E + \chi^{(3)} \cdot E \cdot E \cdot E + \cdots), \tag{2.2}$$

where ε_0 is the permittivity of free space and E is the vector of the applied electric field. $\chi^{(i)}$ is the complex optical susceptibility tensor of the ith rank [185, 212–214]. Parametric oscillation is a three-wave process (Fig. 2.9) and requires materials with a large second-order susceptibility $\chi^{(2)}$.

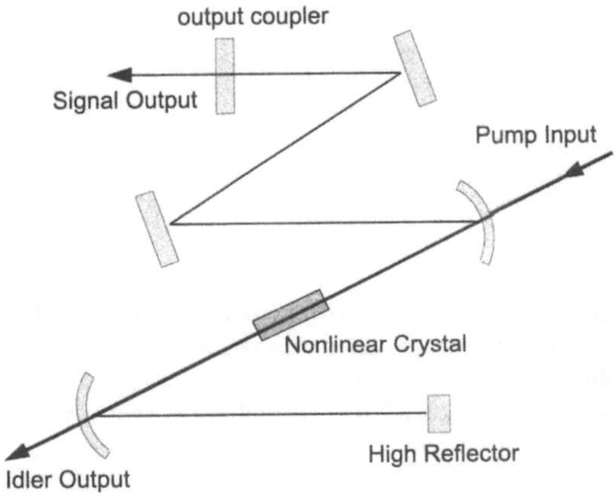

Fig. 2.9. Schematic sketch of an optical parametric oscillator (OPO) cavity. The OPO uses a nonlinear optical crystal to generate new frequencies (called signal and idler) from a coherent input source (pump laser)

In an OPO, energy conservation dictates that

$$\frac{1}{\lambda_{\text{pump}}} = \frac{1}{\lambda_{\text{signal}}} + \frac{1}{\lambda_{\text{idler}}} \tag{2.3}$$

and momentum conservation requires

$$k_{\text{pump}} = k_{\text{signal}} + k_{\text{idler}}. \tag{2.4}$$

By employing birefringent nonlinear crystals, the second condition is satisfied through the matching, which governs the signal and idler output wavelengths.

In a singly resonant OPO, only one of the generated wavelengths is resonated in the cavity. Unlike a laser, an OPO has no storage mechanism. Gain is available only while the pump pulse is present in the nonlinear crystal. Therefore, for continuous wave mode-locked OPOs, a synchronous pumping condition is necessary in which the mode-locked pump laser cavity is precisely matched to the cavity length of the OPO [215–219]. Subpicosecond output pulses require a cavity mismatch below 5 µm. This can be achieved with an actively length-stabilized OPO cavity. Since in case of cavity matching the pump, signal, and idler are all synchronized, with each combination of these wavelengths pump&probe experiments can be performed.

In order to produce wavelengths around 600 nm (as required for the Na_3 B state investigations, see Sect. 3.2.4), a synchronously pumped optical parametric oscillator (Spectra Physics model Opal) (Fig. 2.10) is used, generating

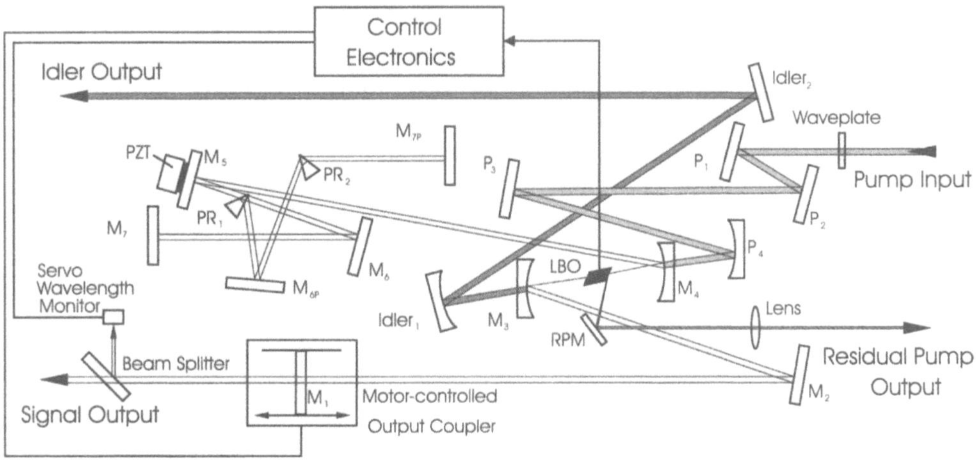

Fig. 2.10. Schematic sketch of Opal cavity (by courtesy of Spectra Physics). LBO nonlinear crystal; P_1 to P_4 pump mirrors; M_1 to M_7 cavity mirrors; $PR_{1,2}$ prisms to control overall cavity dispersion balance in case of 1.3 μm operation; $M_{6P,7P}$ mirrors for 1.3μ m operation; RPM residual pump mirror; PZT piezoelectric transducer

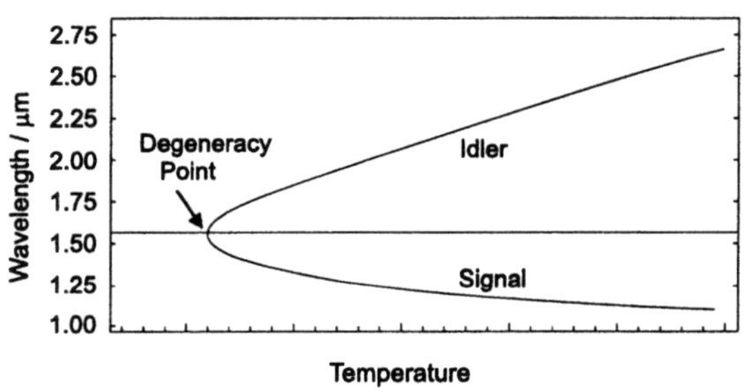

Fig. 2.11. Signal and idler wavelengths of an LBO crystal. In the Opal the wavelengths are tuned by changing the crystal's temperature. The degeneracy point occurs at twice the pump wavelength (by courtesy of Spectra Physics)

a signal wavelength around 1.2 μm. Its signal output is subsequently frequency doubled. The Opal is pumped by a regeneratively mode-locked titanium sapphire laser (Spectra Physics model Tsunami). Matching of the cavity lengths of the Opal and the titanium sapphire laser to better than 5 μm is achieved with a mirror driven by a piezoelectric transducer (PZT) in the Opal cavity.

The Opal uses a temperature-tuned, 90° phase-matched lithium triborate (LBO) crystal as its nonlinear gain medium (see Fig. 2.11). It is pumped at either 775 nm or 810 nm with an average power of 2 W at a repetition rate of 82 MHz; the pump pulse width amounts ∼ 80 fs. The Opal produces two output wavelengths, the main resonated signal output and a less powerful (≈ 50%) idler output. Table 2.1 summarizes the performance obtained.

The 400 mW signal wave of the OPO is subsequently frequency-doubled by a home-built second-harmonic generator. A β-barium borate (BBO) crystal of 100 μm thickness is used as a nonlinear medium [220]. This setup sup-

Table 2.1. Output characteristics of the Opal

Pump wavelength / nm	775	810
average power of signal / mW	400	300
pulse width of signal / fs	< 110	< 120
tuning range of signal / μm	1.10–1.35	1.34–1.60
tuning range of idler / μm	1.80–2.25	1.63–2.00

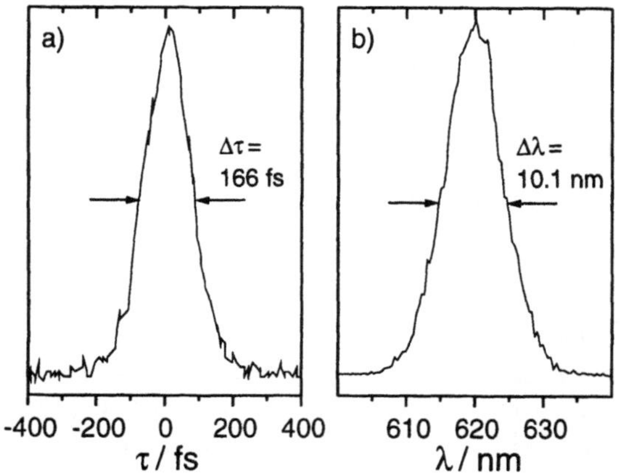

Fig. 2.12. Autocorrelation trace (**a**) and spectrum (**b**) characterizing the frequency-doubled (BBO crystal) femtosecond-Opal. Assuming a sech^2 pulse shape, a pulse width $\Delta t = 110\,\text{fs}$ (FWHM) is measured. The spectrum shows a bandwidth of $|\Delta\bar{\nu}| = (1/\lambda^2)\Delta\lambda = 260\,\text{cm}^{-1}$

plied tunable femtosecond light pulses in the spectral range between 600 and 690 nm. The pulse width was $\Delta t = 110\,\text{fs}$ and the spectral width was estimated to be $|\Delta\bar{\nu}| = (1/\lambda^2)\Delta\lambda = 260\,\text{cm}^{-1}$ (see Fig. 2.12).

Synchronization of Two Mode-Locked Titanium Sapphire Lasers. Besides utilizing OPOs – described above – the use of two independently tunable ultrafast laser sources is possible. This is of special interest while using picosecond laser sources, since the efficiency of nonlinear optical processes is, due to the peak power, much lower than for equivalent femtosecond processes.[10] To achieve the conditions for the pump&probe technique the pulse trains of the two independent lasers have to be synchronized. A successful approach to this problem is described below. For further details on the design of the appropriate stabilization see [50].

Two mode-locked titanium sapphire lasers (Spectra Physics, Tsunami) with 1.4 ps pulse duration and a pulse repetition rate of 80 MHz are synchronized using the master–slave technique [222]. The actual repetition rate of one of the two lasers, the so-called master, serves as a reference oscillator. The second laser, the so-called slave, is matched to the repetition rate of the master. This means that the slave oscillator follows the changes of the master's resonator length. A servo loop is used to realize the alignment procedure.

[10] Recently, the first picosecond OPOs have entered the market [221].

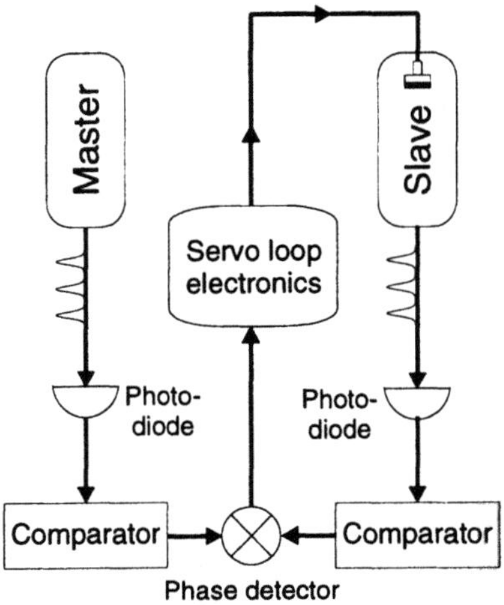

Fig. 2.13. Locking scheme of the master–slave technique used (taken from [223]). The cavity length of the slave resonator is matched to the cavity length of the master resonator by means of a servo loop. See text for further explanation

The locking scheme, shown in Fig. 2.13, demonstrates the principal parts of the signal processing. Two fast PIN photodiodes of 1 GHz bandwidth monitor the actual repetition rates, f_{master} and f_{slave}, of each resonator's output pulse trains. These signals are digitized and compared by a fast phase–frequency detector (800 MHz bandwidth). Any difference in frequency and phase is converted to a proportional error signal. The corresponding voltage is applied to a piezoelectric transducer mounted on one of the intracavity folding mirrors of the slave laser (see Fig. 2.14). The piezo aligns the cavity length of the slave to the cavity length of the master. In the case of perfect overlap of the two pulse trains, there is no further phase error. Then the cavity length of the slave laser is held equal to that of the master. Hence, one maintains the synchronization of the two lasers.

To characterize the quality of the synchronization electronics, the relative timing jitter is the most important quantity. This timing jitter is defined as the temporal walk-off between the pulse trains of the two lasers and was determined by various experimental techniques. First the cross correlation of both pulse trains was determined by second harmonic generation (SHG) in an LBO crystal. In this case both lasers were tuned to the same wavelength. The trace of the cross correlation during a measurement time of one second is shown in Fig. 2.15 a. The cross correlation width $\Delta t_{\mathrm{cross}}$ amounts to 2.4 ps (FWHM). Let the FWHM of the timing jitter be $\Delta t_{\mathrm{jitter}}$. Assuming a Gaussian profile, $\Delta t_{\mathrm{jitter}}$ can be calculated from the known durations (FWHM) of the laser pulses $\Delta t_{\mathrm{master}}$ and $\Delta t_{\mathrm{slave}}$ and the FWHM of the cross correlation $\Delta t_{\mathrm{cross}}$ by

$$\Delta t_{\mathrm{jitter}} = \sqrt{\Delta t_{\mathrm{cross}}^2 - \Delta t_{\mathrm{master}}^2 - \Delta t_{\mathrm{slave}}^2}\,. \tag{2.5}$$

The pulse widths of the master and slave lasers are 1.4 ps. Hence, we have $\Delta t_{\mathrm{jitter}} \leq 1.4$ ps. This small timing jitter allows the study of real-time phe-

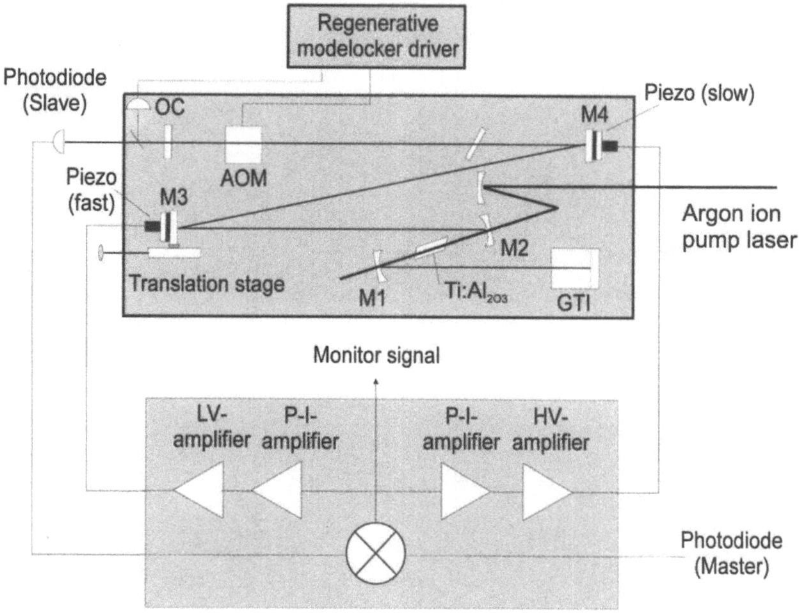

Fig. 2.14. Schematic setup of the regenerative mode-locked titanium sapphire laser (Spectra Physics: Tsunami) with a simplified diagram of the servo electronics used to phase lock the cavity of the slave resonator to that of the master resonator: AOM acusto-optic modulator; OC output coupler; GTI Gires–Tournois interferometer; M1 to M4 high reflectors (taken from [223])

nomena in the picosecond time domain, as successfully performed for the C state of Na_3 (see Sect. 4.1).

Another technique, especially useful to characterize the stabilization electronics more precisely, is the phase noise spectral density $S_J(f)$ [224]. The phase noise spectral density (PNSD) gives information about the amplitude and frequency domain, where most of the noise occurs. The PNSD can be determined by the error signal given as the output (monitor) of the phase–frequency detector (see Fig. 2.14). While the lasers are locked, a Fourier transformation of the monitor signal is calculated to give the PNSD as shown in Fig. 2.15 b. Using the slope of the disperse signal, it is possible to convert the noise voltage to a phase error. Similar values for the relative timing jitter are found by this method, giving a timing jitter less than 1.4 ps.[11]

Autocorrelator. The autocorrelation technique is the most common method used for determining pulse width characteristics on a picosecond and femtosecond timescale. The technique effectively transforms differences in optical

[11] Nowadays similar synchronization units are commercially available using a quite similar technique. The Spectra Physics Lock-to-Clock Tsunami System 3960C is specified with a timing jitter (rms) below 3 ps for the picosecond configuration.

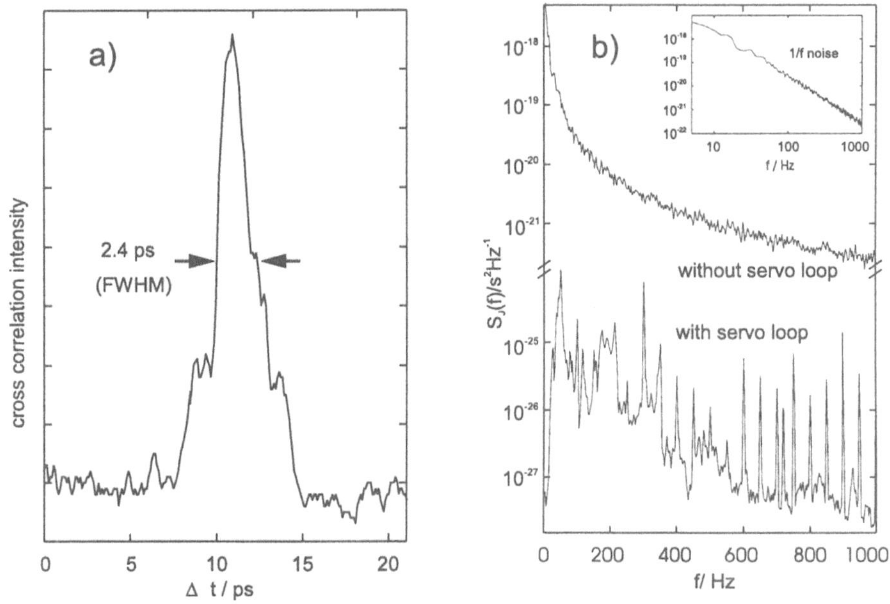

Fig. 2.15. (**a**) Cross correlation trace of the synchronized titanium sapphire lasers (averaging time 1 s) obtained by SHG in an LBO crystal with cross correlation width $\Delta t_{\text{cross}} = 2.4\,\text{ps}$. The pulse widths of the lasers are both 1.4 ps. (**b**) Phase noise spectral density $S_J(f)$ for open and closed servo loop. Note the break in the scale of $S_J(f)$. The figures are taken from [223]

path length into time information, by taking advantage of the fact that the speed of light within a given medium is constant.

The basic optical configuration (see Fig. 2.16) is similar to that of a Michelson interferometer. An incoming pulse train is split into two beams of equal intensity. An adjustable optical delay is imparted to one of the beams and the two beams are recombined within a nonlinear crystal for second-harmonic generation (SHG). The efficiency of the SHG resulting from the interaction of the two beams is proportional to the degree of pulse overlap within the crystal. Monitoring the intensity $I_{2\omega}$ of the UV light as a function of delay time τ between the recombining pulses produces a correlation function directly related to the width of the incoming pulses

$$I(\tau) = \langle I_{2\omega}(t) \rangle \propto 1 + 2G^{(2)}(\tau) \tag{2.6}$$

where $G^{(2)}(\tau)$ is the second-order autocorrelation function of the time-dependent intensity $I^{2\omega}(t)$ given by

$$G^{(2)}(\tau) = \frac{\int\limits_{-\infty}^{\infty} I(t)I(t-\tau)\mathrm{d}t}{\int\limits_{-\infty}^{\infty} I^2(t)\mathrm{d}t}. \tag{2.7}$$

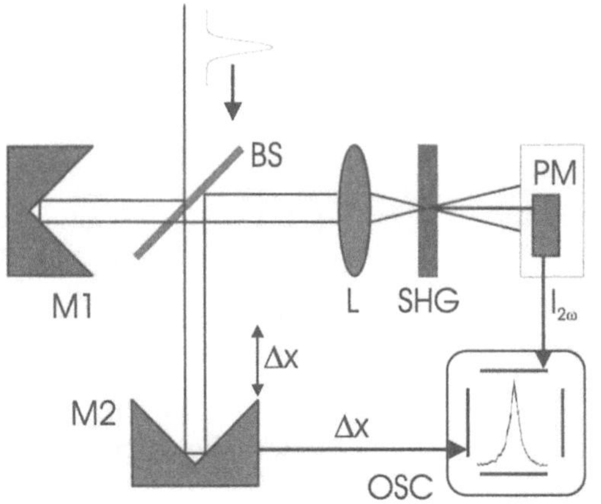

Fig. 2.16. Schematic sketch of an autocorrelator. BS beam splitter; M1, M2 moveable mirrors; L lens; SHG nonlinear crystal for SHG; PM photomultiplier; OSC oscilloscope

Under certain conditions the pulse duration can be calculated from the width $\Delta\tau_A$ of $G^{(2)}(\tau)$. The relation between the pulse width $\Delta\tau$ and the autocorrelation width $\Delta\tau_A$ depends on the shape of the pulse, as shown in Table 2.2. For further details on the autocorrelation technique see e.g. [183, 185, 225, 203].

Table 2.2. Autocorrelation width and spectral bandwidth for different transform-limited pulses taken from [183]. $\Delta\tau$ and $\Delta\tau_A$ are the FWHMs of $I(t)$ and $G^2(\tau)$ respectively. $\Delta\nu$ is the FWHM of the measurable frequency spectrum.

Shape	$I(t)$	$\Delta\tau_A/\Delta\tau$	$\Delta\tau\Delta\nu$
Rectangular	$1(0 \leq t \leq \Delta t)$	1	0.886
Gaussian	$\exp(-(4\ln 2)t^2/\Delta t^2)$	$\sqrt{2}$	0.441
Squared hyperbolic secant	$\text{sech}^2(1.76/\Delta t)$	1.55	0.315
Single exponential	$\exp(-(\ln 2)t/\Delta t)\ (t \geq 0)$	2	0.11

It has to be emphasized that the second-harmonic autocorrelation provides only limited intensity information and no phase information. With decreasing pulse width (e.g. in the sub-50 fs regime) the phase, however, becomes a very important parameter of the laser pulse. Recently, the technique of frequency-resolved optical gating (FROG) has been introduced [226, 227] for the characterization and display of arbitrary femtosecond pulses. FROG overcomes the limitations mentioned above. It can determine the intensity and phase evolution of an arbitrary ultrashort pulse. The FROG technique involves splitting the pulses to be measured into two replicas and crossing them in a nonlinear optical medium. The nonlinear mixing signal is then resolved spectrally as a function of the delay between the two beams. For de-

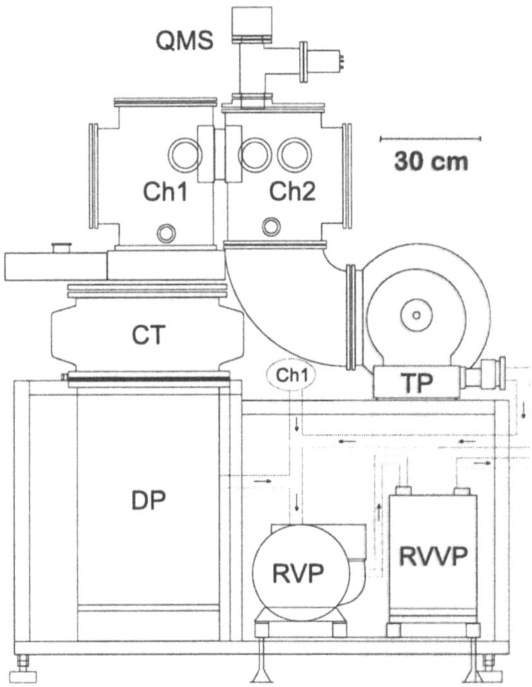

Fig. 2.17. Side elevation of the vacuum setup of the molecular beam machine. Ch1: vacuum chamber to generate the molecular beam by adiabatic expansion. The oven shown in Fig. 2.19 is inside. Ch2: vacuum chamber where the molecules interact with the laser pulses and are detected by a quadrupole mass spectrometer (QMS) or a Langmuir–Taylor detector (Fig. 2.22). Ch1 is pumped by an oil diffusion pump (DP) with a cold trap (CT), Ch2 by a turbomolecular pump (TP). Prevacuum is provided by a Roots (RVP) and a rotary valve vacuum pump (RVVP)

tails on this state-of-the-art technique to characterize ultrashort laser pulses see also [228–237].

The Molecular Beam Machine and the Detection System. The second component of the real-time MPI experiments is a molecular supersonic beam machine [116] with a quadrupole mass spectrometer (QMS), allowing the detection of ionized molecules and clusters with high sensitivity. A side elevation is shown in Fig. 2.17. The production of the molecular beam and the interaction of the laser pulse trains with the molecular beam are performed in a differentially pumped vacuum apparatus consisting of two separate chambers, which are briefly described in the following two paragraphs. A more detailed sketch of the two-chamber system is presented in Fig. 2.18. The production sub-chamber (oven chamber) is pumped by a 3000 ℓ/s oil diffusion pump (Balzers) with a baffle at the flange to the oven chamber to allow a pressure in the chamber of less then 10^{-3} mbar. During the experiments the pressure is typically 5×10^{-4} to 3×10^{-3} bar. In the second chamber a maximum pressure of 10^{-5} mbar is established by a 2200 ℓ/s turbomolecular pump (Balzers).

Generation of the Molecular/Cluster Beam. In the first chamber the molecules and clusters are produced in a seeded supersonic nozzle source. It is perhaps the most intense molecular/cluster beam source available. In this source, alkali metal is vaporized in a hot oven as sketched in Fig. 2.19. The alkali

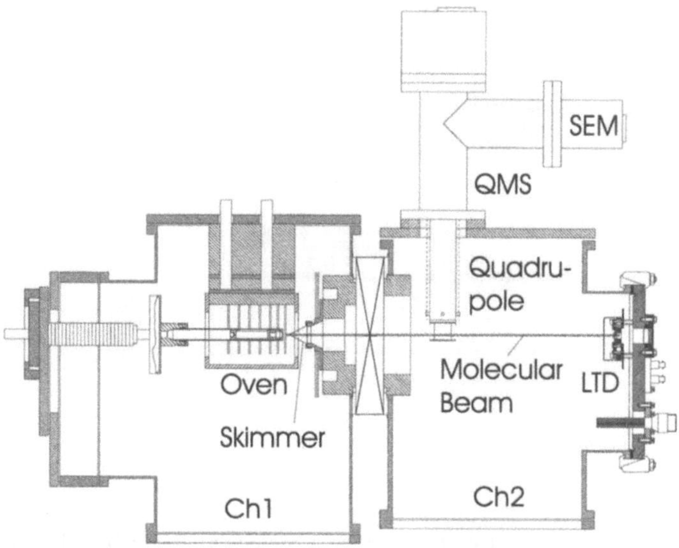

Fig. 2.18. Side elevation of the molecular beam machine's two-chamber setup. Ch1: vacuum chamber to generate the molecular beam by adiabatic expansion. A seeded supersonic beam source is placed here. Further details are shown in Fig. 2.19. Ch2: here, the molecules interact with the laser pulses. Detection is performed by a quadrupole mass spectrometer (QMS) with a 90° ion deflector between the mass filter and secondary electron multiplier (SEM) (Fig. 2.20) or the Langmuir–Taylor detector (Fig. 2.22)

metal vapor is mixed with (i.e. seeded in) an inert carrier gas by pressurizing the oven with inert gas [88, 238–244]. Here, argon is used as the carrier gas. Typically the inert gas pressure is 2 to 7 bar, whereas the metal vapor pressure is in the range of 10 to 100 mbar. The vapor/gas mixture is ejected into a vacuum via a small nozzle, producing a supersonic beam. The expansion into the vacuum proceeds adiabatically and causes dramatic cooling of the mixture [245]. The cooled alkali metal vapor becomes supersaturated, condensing mostly in the form of small aggregates (molecules and clusters). Cluster production continues until the vapor density becomes to low too promote further growth, typically within a few nozzle diameters from the nozzle exit. Likewise, the cooling of the clusters continues until the inert gas density has decreased so much that the flow of the aggregates is molecular rather than hydrodynamic. The mechanism of nucleation and growth of molecules and clusters is discussed in more detail in e.g. [246, 247].

The cooling provided by the gas coexpansion may be adequate to stabilize the aggregates against evaporation, but if it is not, the aggregates may cool even further by evaporating one or more atoms. Evidently, both processes take place. Spectroscopic measurements on alkali dimers and trimers have demonstrated that very low temperatures can be achieved [248]. Rota-

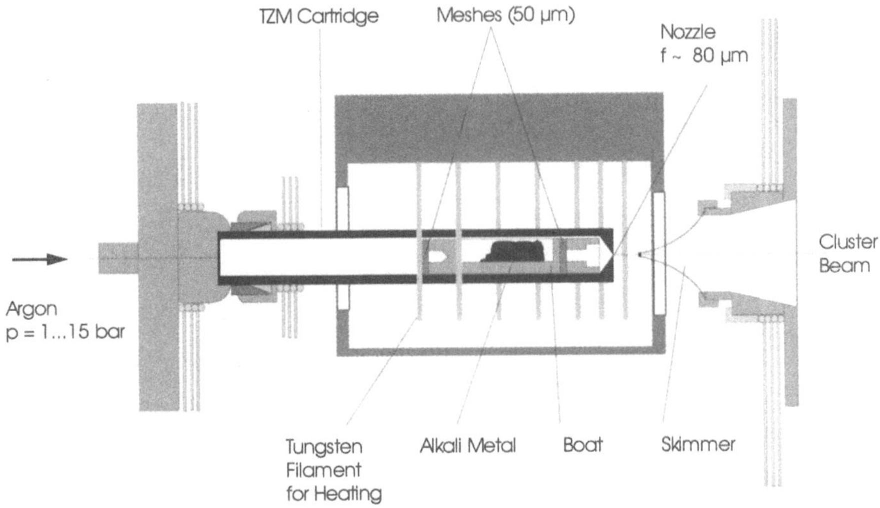

Fig. 2.19. Schematic of the oven used to produce a molecular beam by adiabatic expansion The nozzle diameter is about 80 μm, and the grid size of the mesh is 50 μm. See text for further explanation

tional temperatures of the dimer of 7 K are achieved when applying high gas expansion pressure.

The seeded supersonic nozzle beam used here is shown in Fig. 2.19. This oven consists of a molybdenum alloy tube (TZM: titanium, molybdenum, zirconium) with a cylindrical nozzle 70 μm in diameter, which is filled with about 2 cm^3 of pure alkali metal (sodium or potassium) in a boat. The oven tube is radiatively heated by two separately heatable groups of tungsten filaments, one filament group located around the nozzle region of the oven, the other around the part of the oven tube containing the alkali metal. The currents through the filament groups are adjusted such that the temperatures for the evaporation of the sodium (or potassium) and of the nozzle are as given in Table 2.3. This allows one to establish a temperature gradient, which avoids the nozzle becoming clogged during the supersonic expansion. The alkali vapor is coexpanded with argon (2–7 bar, $p_{max} = 15$ bar) as a carrier gas through the nozzle. This provides a continuous beam of cold alkali molecules with rotational and vibrational temperatures of about 10 K and 50 K respectively [249].

Detection of the Molecular/Cluster Beam. The molecular beam, mainly containing monomers, dimers, and trimers, but also larger clusters, enters the second chamber (interaction chamber) through a 1 mm diameter skimmer (Beam Dynamics Co.) to get rid of the outer part of the molecular beam. To allow mass-selected analysis of the molecular beam a quadrupole mass spectrometer with its axis perpendicular to the molecular beam is installed in this chamber.

Table 2.3. Typical parameters of the molecular beam oven during measurement with sodium and potassium aggregates

	Sodium	Potassium
Oven temperature / K	850	800
Nozzle temperature /K	950	900
Argon pressure / bar	2–7	2–7
Beam time / h	≤24	≤14

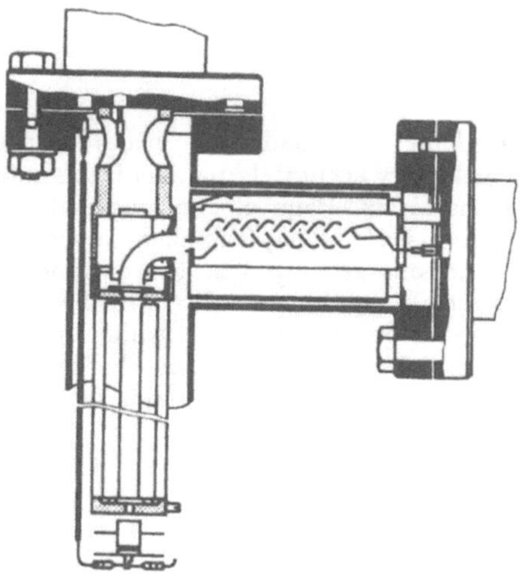

Fig. 2.20. Schematic of the quadrupole mass spectrometer (QMS) with 90° ion deflector between the mass filter and secondary electron multiplier

As seen in Figs. 2.20 and 2.21, the QMS employed consists of three main parts. A quadrupole mass filter is followed by a 90° ion deflector which pilots the ions into a secondary electron multiplier.

The electric quadrupole mass filter, developed by W. Paul and coworkers [250, 251], operates on the principle that ion trajectories in a two-dimensional quadrupole field are stable if the field has an ac component $E_V(\omega)$ superimposed on a dc component E_U with appropriate amplitudes and frequency (ω). Using the equation of motion for an ion in these fields, it can be shown that with correctly adjusted fields the trajectories of singly charged ions are stable only within a narrow mass range. The condition for critical stability is obtained when $E_U/E_V = 0.168$ [251]. The resolution of the QMS is determined by how close to the critical stability the quadrupole mass filter is operated. A resolution of $m/\Delta m \sim 10^2$ to 10^3 is typical. Behind the mass filter the mass-separated ions are deflected and directed to an SEM. The SEM output signal gives a measure for the number of ions with the chosen mass-to-charge ratio (m/e).

The stability of the molecular beam is controlled on-line by a home-built Langmuir–Taylor detector (LTD) [252]. The LTD can be regarded as one of the most effective and widely used detectors for small alkali particles in the

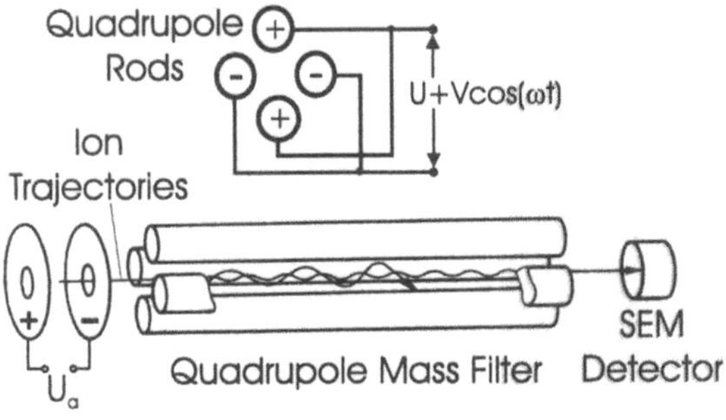

Fig. 2.21. Schematic of a quadrupole mass filter. The quadrupole pieces are precisely machined cylindrical rods, which are held very accurately parallel in a square configuration (see Fig. 2.20) with ceramic insulators. Pairs of opposing rods are electrically connected. A time-dependent potential $U + V(\omega)$ is applied to one pair of rods and $-(U + V(\omega))$ to the other. The ions generated at the entrance of the quadrupole rods are accelerated (voltage U_a) into the mass filter. In the high-frequency electric quadrupole field the ions can be separated according to their mass-to-charge (m/e) ratio. Hence, only cluster ions with a mass-to-charge ratio corresponding to the applied ac and dc voltages will pass through this filter and reach the SEM detector

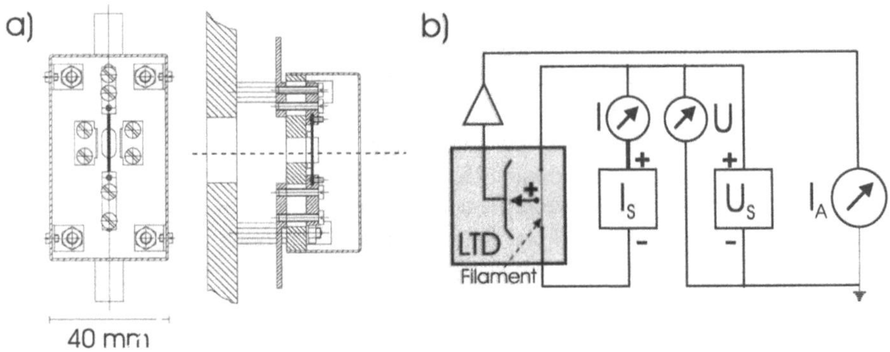

Fig. 2.22. Langmuir–Taylor detector. **(a)** Side and top elevation. The detector's filament is made of rhenium [252]. **(b)** Schematic of the electronic measuring circuit. The filament is heated by a stabilized current I_S and set to a positive potential by a second highly stabilized voltage U_S. The preamplified anode current I_A is detected by a lock-in technique or by a picoammeter

gas phase [253]. This surface ionization detector was first developed by Taylor [254, 255] in 1929 on the basis of Langmuir's observation [256–259] that every Cs atom which struck a heated wolfram wire came off as a positive ion. Once the atom is ionized it can be measured in any of several different ways. In Fig. 2.22 the home-built LTD and its installation in an electronic measuring circuit are sketched. A rhenium wire is used as a filament. The LTD allows the estimation of the molecular beam intensity and profile. Its intensity can be used to roughly normalize the real-time spectrum.

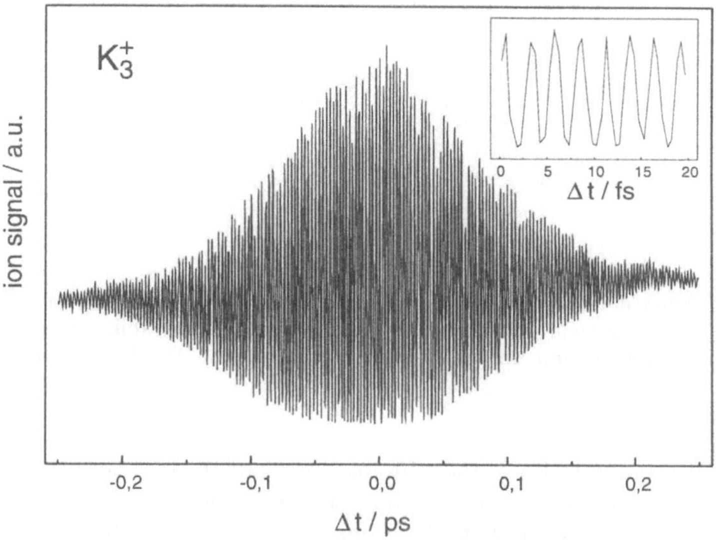

Fig. 2.23. Interferometric autocorrelation trace directly measured at the molecular beam. The 3PI signal of K_3 is detected while exciting with 90 fs laser pulses at $\lambda = 800$ nm. The inset nicely depicts the fast 2.67 fs oscillation (taken from [260])

Method and Technique of the Measuring Process. The laser beam is split and recombined collinearly with parallel polarization in a Michelson-like arrangement. The length of one of the Michelson branches is controlled by a computer-driven DC-motor translation stage [261]. The position of the translation stage is read out by an optical encoder. An ultrashort (probe) laser pulse, passing through this arm of the Michelson arrangement, is compared to the fixed branch shifted by their length difference Δx. Therefore, it is optically delayed in time with respect to the pump pulse passing through the other Michelson branch. The delay time Δt between the pump and the probe pulse is given by $\Delta t = 2\Delta x / c_\lambda$, where c_λ is the speed of light in air for the wavelength λ used. The accuracy of the delay time Δt is about 1 fs, and the minimal step width amounts to 0.3 fs.

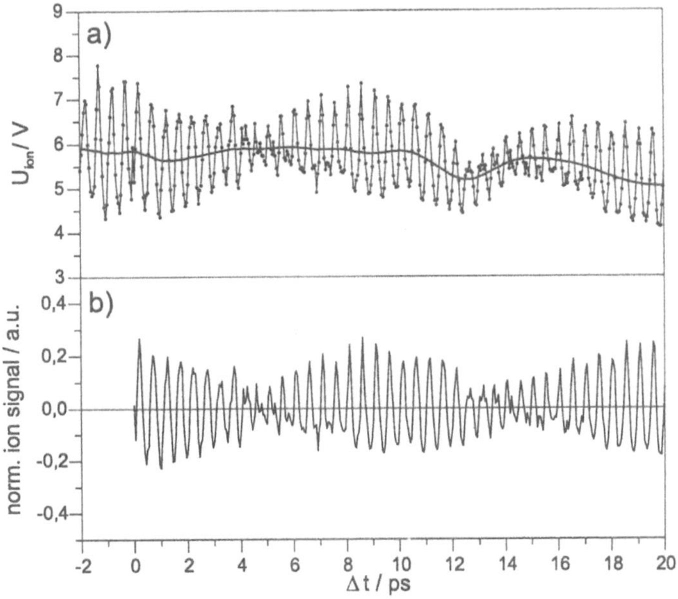

Fig. 2.24. (**a**) Typical output signal of the quadrupole mass spectrometer. The fluctuations of the real-time spectrum are mainly due to instabilities of the molecular beam. (**b**) Real-time spectrum after normalization. The figure is taken from [262]

It is of great interest to know the time resolution of the real-time experiment at the actual spot, where the ultrafast dynamics of the molecules and clusters are investigated. Owing to GVD, femtosecond laser pulses tend to broaden in the time domain while passing through the glass used at several places (e.g. focusing lens and entrance window to the vacuum chamber) in the setup. Therefore, it is important to measure the pulse duration directly, in the interaction zone of the laser pulses and species under study. Figure 2.23 shows an interferometric autocorrelation trace detected by real-time three-photon ionization (3PI) of K_3 molecules excited at $\lambda = 800$ nm. The pulse width is estimated to be 92 fs. The inset neatly reveals an oscillation of 2.67 fs period, which is due to the interference of the pump and probe pulses with a central frequency of 3.74×10^{14} Hz ($\hat{=}$800 nm).

The laser beams are weakly focused on the molecular beam by a 400 mm quartz lens, and each pulse reaches (in the femtosecond experiments) a peak power of about 0.5 GWcm^{-2}. Photoionized molecules and clusters are guided by electrical lenses into the axis of a quadrupole mass spectrometer (Balzers QMG 420). The ions are mass-selected with a resolution $m/\Delta m > 200$, which is sufficient to distinguish between, for example, isotopic species $^{39,39}K_2$ and $^{39,41}K_2$ (see Sect. 3.1.3). The transmitted ions are collected by a secondary electron multiplier and amplified. This detection setup works linearly up to pressures of 10^{-5} mbar. Typical pressures in the oven chamber and the interaction chamber are 5×10^{-4} mbar and 7×10^{-6} mbar respectively.

The ion signal produced is continuously recorded as a function of the delay time Δt between the pump and the probe pulse. The time to record a real-time spectrum up to 200 ps is about two hours, a time over which the molecular beam is not a priori stable. While in our experiments the intensity of the laser was stable within 3%, the LTD-controlled intensity of the molecular beam varied over a range of about ±20%. Therefore, the transient data were normalized to obtain data points oscillating around the zeroline to perform a correct Fourier analysis (see Fig. 2.24).

2.1.2 Setup for the Real-Time Charge Reversal (NeNePo) Experiments

As introduced in Chap. 1, a real-time 'NeNePo' experiment on molecules or clusters starts with a photodetachment process followed by a photoionization step, each induced by an ultrashort laser pulse (see Fig. 2.25). In Chap. 5 the first, fascinating results measured with this technique are presented. The setup for this high-performance, femtosecond 'NeNePo' charge reversal experiment is shown in Fig. 2.26. The special features of the laser system employed are then given, followed by a few fundamental details on generation, storage, and detection of the investigated particles.

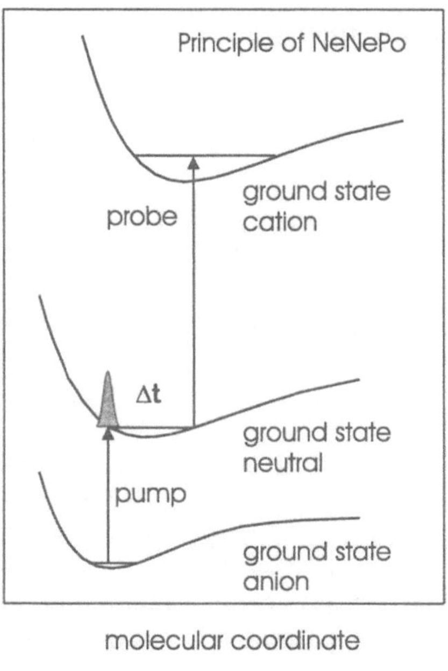

Fig. 2.25. Principle of the NeNePo process. Beginning at the anion's potential-energy surface, an ultrashort pump pulse detaches an electron and prepares a wave packet in the neutral. After a certain delay time Δt a probe pulse photoionizes the neutral. The method is named 'NeNePo', since the overall process is a **N**egative-to-**N**eutral-to-**Po**sitive charge reversal transition

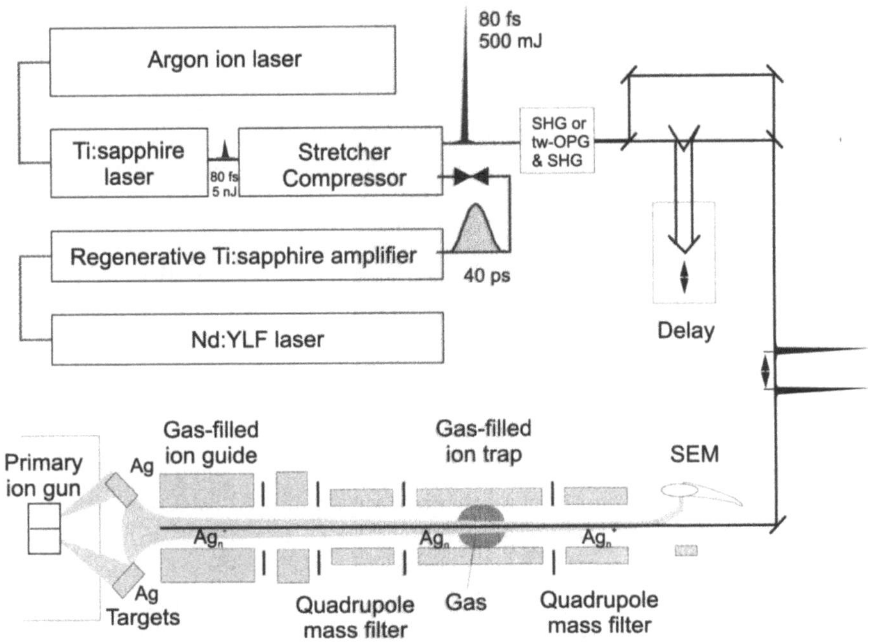

Fig. 2.26. Experimental setup for the NeNePo investigations (taken from [263]). For explanation see text

Ultrashort Laser System. To analyze the time evolution of the neutral trimer's configuration a laser system consisting of a titanium sapphire oscillator (Spectra Physics Tsunami) which is pumped by a 12 W argon ion laser, and a Nd:YLF-pumped regenerative amplifier system (Quantronix 4800 titanium sapphire RGA) is employed [175].

Regenerative Amplifiers. These allow maximum energy extraction using a modicum of hardware, making them extremely well-suited for seeding with low-energy pulses. Solid-state regenerative amplifiers come in two classes: cw-pumped and pulse-pumped. The former offer higher repetition rate, the latter higher energy. The Nd:YLF pump laser used here is Q-switched, and hence belongs to the second class. At 1 kHz the laser provides more than 15 mJ pump energy.

The amplifier system is composed of three logic subunits, a femtosecond pulse stretcher (Fig. 2.27), the titanium sapphire regenerative amplifier (Fig. 2.28) and a femtosecond pulse compressor (Fig. 2.29). To exploit the storage capability of titanium sapphire the system uses chirped-pulse amplification [264–266]. For this a seed pulse (80 fs, 5 nJ) of the titanium sapphire oscillator is stretched to about 500 ps and then amplified. The stretched pulse can be safely amplified because the peak power has been significantly lowered, thereby avoiding any nonlinear effects that could result in damage to the

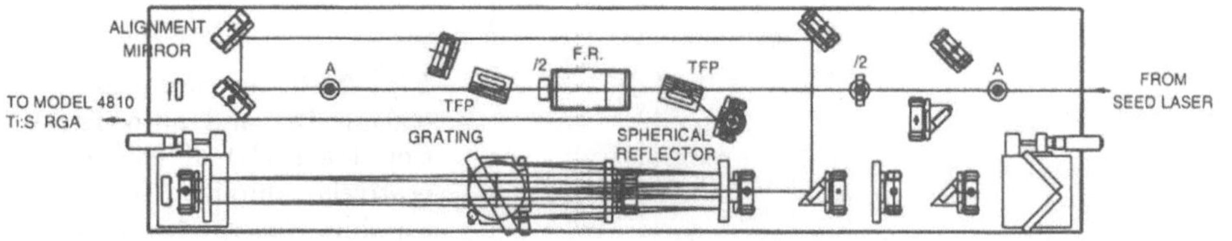

Fig. 2.27. Top view of the femtosecond pulse stretcher (Quantronix model 4822) and laser beam tracing. The pulse from the seed laser is stretched about 5000 times with the use of a diffractive grating mounted at Littrow incidence to reduce aberrations. To prevent optical feedback to the seed laser a Faraday rotator (FR) is used for optical isolation. A aperture; TFP thin-film polarizer (by courtesy of Quantronix)

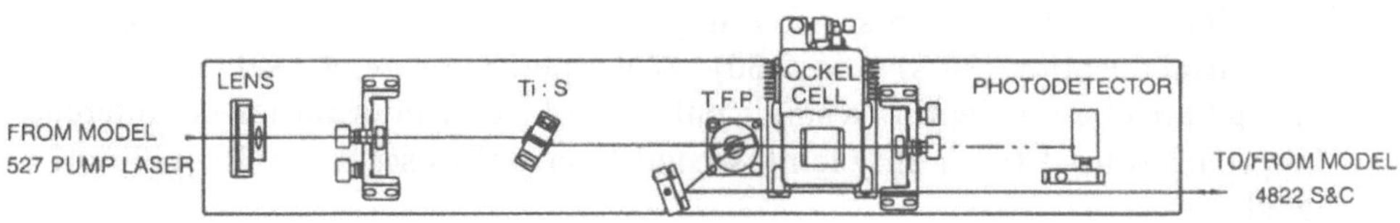

Fig. 2.28. Top view of the amplifier cavity (Quantronix model RGA 4810). The KD*P (KD_2PO_4) Pockels cell (PC, Medox Electro-Optics) has a dual function. First, it frustrates cavity lasing except for a brief time period when the seed pulse is accepted for amplification. Secondly, in combination with the thin-film polarizer (TFP), it also controls the injection of the seed pulse and extraction of the amplified pulse. The Pockels cell operates at 1 kHz. A photodetector is provided for intracavity power monitoring, which is necessary for adjusting the two time delays of the Pockels cell driver (by courtesy of Quantronix)

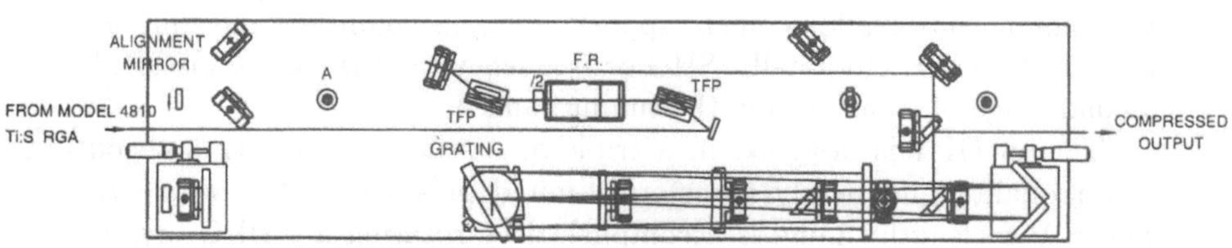

Fig. 2.29. Top view of the femtosecond pulse compressor (Quantronix model 4822) and laser beam tracing. The compressor is built up in the same housing as the stretcher and the laser pulses run a similar way back, using the optical components of the pulse stretcher, especially the diffractive grating. By means of the Faraday rotator (FR) the polarization of the light pulses is turned 90° (45° for each through run) and hence, at the latter thin-film polarizer (TFP) the pulses are now reflected. A aperture (by courtesy of Quantronix)

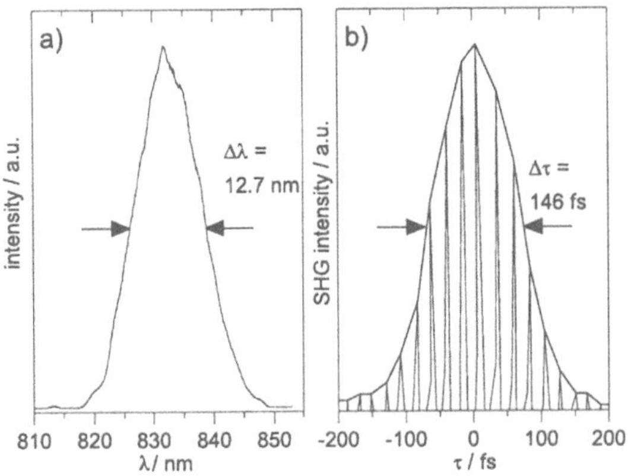

Fig. 2.30. Spectrum at $\lambda = 830\,\text{nm}$ **(a)** and autocorrelation trace **(b)** of the regeneratively amplified titanium sapphire laser. Assuming a sech^2 pulse shape, a pulse width $\Delta t = 94\,\text{fs}$ (FWHM) is measured. The spectrum reveals a bandwidth $|\Delta\bar{\nu}| = (1/\lambda^2)\Delta\lambda = 183\,\text{cm}^{-1}$. With these data the pulse is 1.6 times bandwidth-limited (see Table 2.2) (taken from [262])

amplifier or its components. After amplification, the pulse is recompressed to its initial duration (80 fs) (Fig. 2.30). At a repetition rate of 1 kHz energies of 500 μJ are obtained within a single pulse. For further information on different approaches used to amplify femtosecond laser pulses see e.g. [176, 177, 267–269].

Optical Frequency Conversion. The amplifier system is optimized for a seed wavelength within the tuning range of 770 to 830 nm. To enable further tuning ranges, frequency doubling and optical parametric generation (OPG) are utilized. In a BBO crystal the amplified pulses can be efficiently ($\approx$ 40%) frequency-doubled. Besides this, a travelling-wave (tw) optical parametric amplifier (OPA) using superfluorescence is employed. This OPA converts the amplified laser pulses of fixed wavelength into a broadly tunable radiation. Its principle of operation is based on high-gain amplification of quantum noise in a transparent nonlinear crystal pumped by intense ultrashort pulses. In contrast to the synchronously pumped OPO (see Sect. 2.1.1), the travelling-wave configuration allows one to apply single-pulse pumping and thus achieve high peak power. Optionally, SHG or frequency mixing can be installed as a second stage of transforming the tuning range.

The tw-OPA is designed in a triple-pass scheme, ensuring high output beam quality and nearly transform-limited pulses over the whole tuning range. Wavelength tuning is accomplished by rotating a β-BBO crystal. To protect the crystal against humidity it is heated up to 40–50° C. Figure 2.31 presents a sketch of the tw-OPA.

Generation, Storage, and Detection of Small Mass-Selected Silver Molecules and Clusters. In addition to the source of femtosecond pulses for the 'NeNePo' investigations, negatively charged molecules and clusters (here silver) are necessary. The vacuum system is described in detail in [271].

Generation of Mass-Selected Anions. The negatively charged molecules and clusters are generated in a sputtering ion source by bombarding targets of

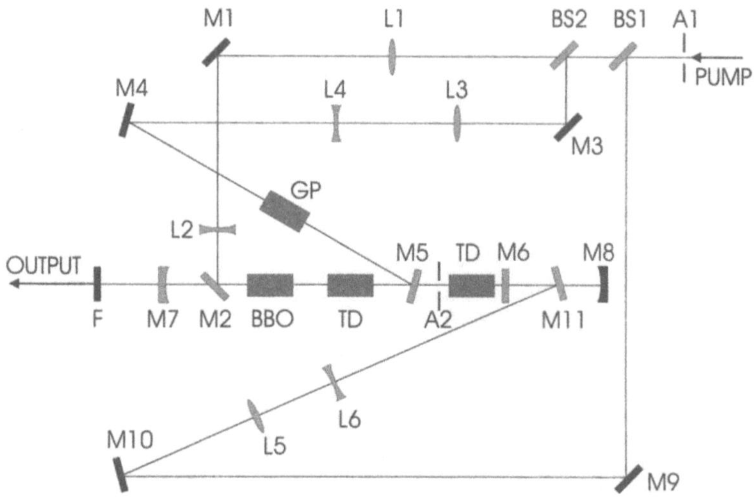

Fig. 2.31. Top view of the optical scheme of the travelling-wave optical parametric amplifier (tw-OPA). $A_{1,2}$ apertures; BBO β-barium borate crystal; $BS_{1,2}$ beam splitters, F output window; GP generator of parametric superfluorescence; $L_{1...6}$ lenses; $M_{1...11}$ mirrors; TD temporal delay (taken from [270])

elemental silver with fast Xe^+ ions [272–274]. The sputtering ion source is especially designed to yield a high flux of well-thermalized cluster ions. The xenon cations are generated in a cold reflex discharge ion source (CORDIS) [271, 275, 276]. Most of the sputter secondary particles are neutrals. Depending on the sputtering efficiency, however, a few nanoamps of ions, mostly silver cations, are generated. Vaporizing cesium metal onto the targets lowers the work function of the silver surface and enables a drastic increase of the yield of the desired anions [277, 278].

An ion lens collects the generated anions and leads them into a first quadrupole ion guide of large crosssection (see Fig. 2.32). Here, the ions are cooled and moderated by collisions with a background gas of He ($p_{He} \approx 10^{-2}$ mbar) [279]. Thereby internal energy is removed from the clusters, and the volume of the ion cloud both in real space and in momentum space is substantially compressed. This matches the emittance of the sputtering ion source to the acceptance of the subsequent quadrupole mass spectrometer. The anions are then approximately at room temperature. Next the anions pass through a quadrupole mass filter, where the species of interest can be selected. A beam of well-thermalized mass-selected metal cluster ions is created by this method. The final cluster intensity ranges from some nanoamperes in the case of the small clusters (e.g. Ag_3^-) to a few hundred picoamperes for the larger clusters such as Ag_9^-.

Storage of Mass-Selected Neutral Molecules and Clusters. In order to increase the density of the anions, they are accumulated in a linear quadrupole ion

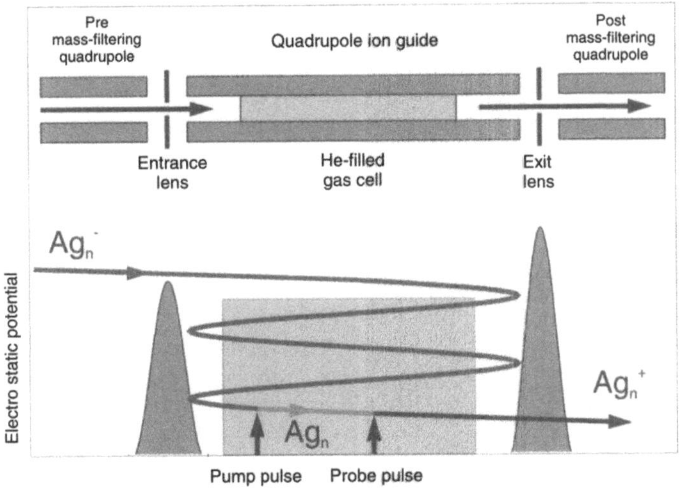

Fig. 2.32. Schematic of the ion trap used for the real-time NeNePo experiments. The first quadrupole prepares a beam of mass-selected negative ions (here Ag_n^-). These are stored in the central quadrupole ion trap. Here, detachment (pump) and ionization (probe) laser pulses interact with the stored molecules or clusters. The positively charged ions are subsequently accelerated to a third quadrupole, where again mass filtering can be carried out. The distance the Ag_n cluster runs on the trajectory between pump and probe pulses is greatly exaggerated. Over a time of a few picoseconds the cluster stays nearly in the same position, since its velocity is rather small

guide, which is filled with helium gas ($\sim 10^{-3}$ mbar). This ion guide operates as an ion trap [280] and leads to further thermalization. For this, the entrance lens of the ion guide is kept at a potential slightly below the kinetic energy of the ions, while the exit lens is at a higher potential. Hence, the anions can enter the ion guide, but are reflected back at the exit lens. Traveling through the gas cell, the ions lose kinetic energy by collisions and are no longer able to escape via the entrance lens. The number of stored ions can be monitored by pulsing the exit lens open and recording the magnitude of the ion-current pulse. The residence time as negative ions can be minutes. When the photodetachment is carried out, the negative ions typically reside in the trap for about 100 ms. During this time, they undergo more then 1000 collisions with the buffer gas. The ion density in the trap is limited by space-charge effects to about 10^7 ions per cm^3, independent of the cluster size. Without a detachment laser beam, more than 10^8 ions can be stored in the trap (Fig. 2.33). With a detachment laser, an equilibrium between the continuous filling and the depletion of the trap by the detachment process is reached. In this case, an ensemble of about 10^6 mass-selected cations is stored and can interact with the subsequent detachment and ionization pulses.

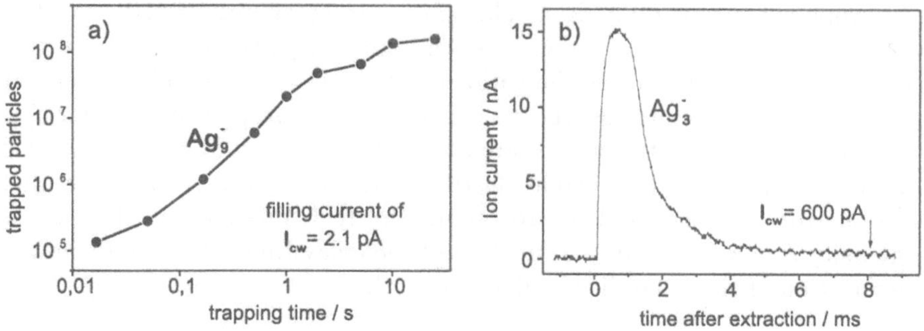

Fig. 2.33. Time characteristics for filling and extracting the storage trap (taken from [271]). **(a)** Number of trapped Ag_9 cluster ions for a given current of filling as a function of time. **(b)** The trap is filled for 200 ms with Ag_3. At $t = 0$ the exit lense is opened. Within 5 ms the ion current decreases to the cw value of 600 pA

Real-Time Detection of Laser-Pulse-Induced Molecular Dynamics. To study silver aggregates the amplified laser pulses are, for example, efficiently ($\approx$ 40%) frequency-doubled in a BBO crystal. The second harmonic is split into pump and probe pulses, with the probe pulse delayed with respect to the pump pulse by a computer-controlled translation stage. The pump and probe laser beams are imaged into the trap collinearly with the ion trajectories and overlap throughout the whole length of the trap (see Fig. 2.26). The electrons of the stored cluster anions are detached by the pump pulse and, subsequently, after a certain delay time Δt, the newly created neutrals are ionized by the probe pulse. This is achieved by nonresonant TPI rather than by REMPI [281, 282]. Figure 2.34 presents typical mass spectra for silver anions and cations.

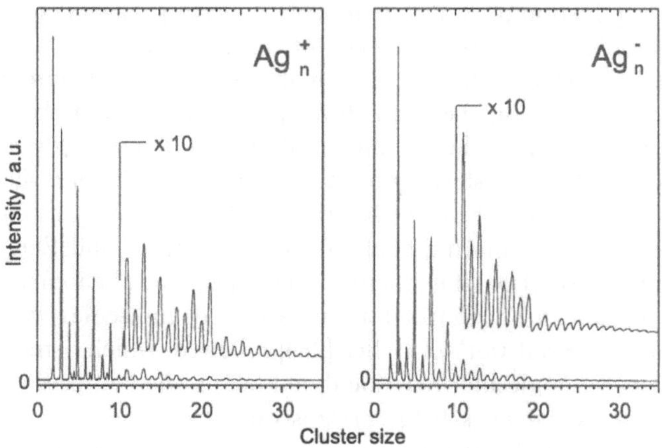

Fig. 2.34. Mass spectra of Ag_n cations and anions (taken from [271])

Applying the tw-OPA, it should be easy to carry out REMPI investigations in the near future. Now, however, the trap acts no more as a trap, but, preferably, accelerates the cations to the ion detector. The cations are extracted by the exit lens and pass a second quadrupole mass filter. The once-more mass-selected cations are then detected by an SEM detector. This allows the detection of possible cluster fragments during or after the interaction with the laser pulses. To obtain the real-time spectra the cation's intensity is now detected as a function of the delay time Δt. A more detailed description of the beam line is given in [271].

2.2 A Few Basics on the Theoretical Approach to Real-Time Spectroscopy

The experiments presented here, require in some cases sophisticated theoretical approaches. The calculations of the wave packet propagation (Sect. 2.2.1) observed for excited dimers and trimers (Chap. 3) are of this special kind. The photodissociation phenomena (Chap. 4), however, can be treated by two rather easy approaches, as described in Sect. 2.2.2. The essentials of thefirst theoretical approaches to interpretion of the results on ultrafast structural relaxation (Chap. 5) obtained by the 'NeNePo' technique are summarized in Sect. 2.2.3.

2.2.1 Theoretical Analysis of Wave Packet Propagation

Recently, two excellent reviews of theoretical approaches to wave packet dynamics with a focus on femtochemistry have been published. In 1995 Garraway and Suominen gave a review of the basic theory and simple properties of wave packets, the application of two-state models to wave packet dynamics, and wave packet processes in the presence of light [8]. In 1998 Manz presented a brilliant historical survey of molecular wave packet dynamics [7]. He starts with an overview on the early development of the underlying theoretical methods and concepts, the fundamental laws, and some general properties of wave packets. Via the 'sleeping beauty' era (1930–1965) of rather little direct progress in the theory of molecular wave packet dynamics, he enters the 'renaissance' of this fascinating field, started by Jortner and Berry [283] and others with the prediction of explicitly time-dependent phenomena, e.g. quantum beats, which can be regarded as signatures of molecular wave packet dynamics. The important work of Heller [284] followed, which demonstrated the relation between molecular reaction dynamics on the femtosecond timescale and continuous wave weak-field spectroscopy. As Manz shows in his contribution, in the early 1980s the 'Sturm und Drang' era produced several leading groups such as Feit and Fleck [285, 286] and D. and R. Kosloff [287, 288] who introduced new and powerful numerical techniques for the propagation of wave packets (fast Fourier transform propagation method). These theoretical approaches then led straightforward to the concept of controlled molecular dynamics pioneered by Paramonov and Manz [289, 290], Tannor, R. Kosloff, and Rice, [60, 61] and Brumer and Shapiro [59].

With this short overview of molecular wave packet dynamics in mind, the wave packet propagation in small prototype molecules will now be examined more concretely. Different theoretical methods exist to describe ultrafast multiphoton ionization processes in diatomics and have been discussed in detail in previous work [3, 33–35, 291–294]. The theoretical approach used here is adapted specially to the presented 3d problems and was improved in Manz's group, mainly by de Vivie-Riedle [295] assisted by Reischl [81]. A few special features of their theoretical ansatz applied to the pump&probe experiments carried out on the model molecules K_2 and Na_3 are now briefly summarized.

The starting point of their approach is the Hamiltonian H_{mol} of the molecular system, which is defined as the sum of the operators for the kinetic energy of the nucleus $T_{\text{k}} = P^2/(2m)$ and of the electrons T_{el} and the potential energy V of the system:

$$H_{\text{mol}} = T_{\text{k}}(Q) + T_{\text{el}}(r) + V(Q,r), \tag{2.8}$$

where Q and r are respectively the nuclear and electronic space coordinates of the system. Within the Born–Oppenheimer approximation the electronic part of the Schrödinger equation

$$H_{\text{mol}}\Psi = E\Psi \tag{2.9}$$

and the equation for the nuclear motion can be solved separately. Different ab initio methods based on the Hartree–Fock formalism are used to adequately solve the electronic Schrödinger equation for the special dimer and trimer systems investigated in this book. This enables the calculation of the relevant potential-energy surfaces of the molecular systems. To describe the dynamics of the quantum-mechanical systems the numerical solution of the time-dependent Schrödinger equation

$$\mathrm{i}\hbar\frac{\partial}{\partial t}\Psi(Q,r,t) = H_{\text{mol}}\Psi(Q,r,t) \tag{2.10}$$

has to be determined. To define the initial state $\Psi(t=0) = \Psi_{1\dots n}$ of an n-dimensional molecular system, first the time-independent Schrödinger equation

$$(T_{\text{k}} + V)\Psi_{1\dots n}(Q) = E_{1\dots n}\Psi_{1\dots n}(Q) \tag{2.11}$$

is solved for the nuclear motion wi*h $Q = (Q_1 \dots Q_n)$. For the potassium dimer (Sect. 3.1) propagation in imaginary time [296] and for the sodium trimer (Sect. 3.2) the 'discrete variable representation' (DVR) [297] are employed. The obtained Hamilton matrix is diagonalized by direct mathematical techniques [298].

As a result of the interaction with ultrashort laser pulses, molecular wave packets are generated on one or several PESs. The dynamics of the molecule initiated by the time-dependent interaction with the light pulse are then described by the Schrödinger equation in matrix representation:

$$i\hbar\frac{\partial}{\partial t}\begin{pmatrix}\Psi_1\\ \vdots\\ \Psi_n\end{pmatrix}=\begin{pmatrix}H_{11}&\cdots&H_{1n}\\ \vdots&\ddots&\vdots\\ H_{n1}&\cdots&H_{nn}\end{pmatrix}\cdot\begin{pmatrix}\Psi_1\\ \vdots\\ \Psi_n\end{pmatrix}. \tag{2.12}$$

n is given by the number of PESs φ_j coupled by the studied multiphoton process. $\Psi_j(Q,t)$ represents the total wave function in the electronic state $\varphi_j(r,Q)$. Equation (2.12) is solved without any further approximations, and within the model all multiphoton processes are included.

The time-dependent Hamiltonian is separated into a time-independent part (diagonal elements H_{ii} of the Hamilton matrix)

$$H_{ii}=T_{\mathrm{k}}+H_{\mathrm{el}}+V_{ii}, \tag{2.13}$$

and non diagonal elements $V_{ij}(i\neq j)$. The time-dependent part is given by the off-diagonal elements $H_{ij}(i\neq j)$ of the Hamilton matrix

$$H_{ij}(t)=-\mu_{ij}E(t). \tag{2.14}$$

H_{ii} describes the time-independent interaction of the coupled PES and V_{ii} is the PES of the molecule in the absence of any laser fields. H_{ij} represents the interaction of the molecule with the laser pulse. The μ_{ij} are the dipole transition moments and $E(t)$ the time-dependent electromagnetic field generated by the superposition

$$E(t)=E_{\mathrm{pump}}(t)+E_{\mathrm{probe}}(t) \tag{2.15}$$

of the ultrashort pump and probe laser pulses. Owing to the coherence and high density of photons within the laser pulses $E(t)$ can be described classically and is given in the dipole approximation by the ansatz

$$E(t)=E_0 s(t)\cos(\omega t). \tag{2.16}$$

The amplitude E_0 of the electromagnetic field is determined by

$$E_0=\sqrt{\frac{2I}{\varepsilon_0 c}} \tag{2.17}$$

for a laser pulse of intensity I; ε_0 is the dielectric constant and c the velocity of light. The function $s(t)$ represents the shape of the laser pulse and ω is the central frequency of the laser pulse.

The basic technique used to propagate the wave packet in the spatial domain is the fast Fourier transform method [287, 288, 299, 300]. The time-dependent Schrödinger equation is solved numerically, employing the 'second-order differencing approach' [299, 301]. In this approach the wave function at $t'=t+\delta t$ is constructed recursively from the wave functions at t and $t''=t-\delta t$. The operator including the potential energy is applied in phase space and that of the kinetic energy in momentum space. Therefore, for each temporal step the wave function has to be Fourier transformed back and forth, which is performed by the fast Fourier transform algorithm [299].

Owing to the ejection of an electron with energy E, a continuum is superimposed on the molecule's ion state. Energy conservation in the global system of molecule plus laser field restricts the total region of the continuum to an interval $[0, \varepsilon]$. Different methods of discretizing the electronic continua are employed in theoretical studies of ultrafast ionization processes [32, 302]. For the investigations on K_2 and Na_3 the continuum is simulated by discretizing the corresponding energy range by a sufficient number N of electronic states [303–305].

2.2.2 Simple Decay Models for Ultrafast Photodissociation

The availability of powerful lasers providing ultrashort pulses in a wide range of frequencies has stimulated – besides rapid development of experimental techniques – the theorists to develop sophisticated methods to treat photodissociation processes, at least for small molecules, in an essentially exact quantum mechanical way. The marvellous book by Schinke [122] gives an excellent overview of the state-of-the-art. However, for clusters only very simple models have been used up to now to analyze the real-time photodissociation.

In this section a rather simple energy-level model is first briefly described, which enables a rough analysis of real-time decay processes obtained by MPI. It has been successfully applied to the investigations of the Na_3 state. To use the model for the larger alkali clusters, however, it has to be modified. Both models can be regarded as a first approximation to estimate the time constants of the induced photodissociation process. It has to be stated that neither of the two models takes into account the dynamics of wave packets prepared on the repulsive PES.

Simple Energy-Level Model. Figure 2.35 sketches a simplified case of TPI spectroscopy of a dissociative electronic state. Fragmentation with a probability $1/\tau_{\text{frag}}$ is regarded as the only relaxation channel. In a real-time TPI experiment, first an ultrashort pump pulse transfers an ensemble of e.g. molecules or clusters to an excited state (2.18a). Then, either the excited state (∗) can dissociate and two fragment products are found (2.18b), or by absorbing a probe photon ($E_{\text{probe}} = h\nu_{\text{probe}}$), more or less delayed in time, the cluster can be ionized (2.18c).

$$\mathrm{M}_n + h\nu_{\text{pump}} \longrightarrow \mathrm{M}_n^* \tag{2.18a}$$

$$\mathrm{M}_n^* \longrightarrow \mathrm{M}_{n-m}^* + \mathrm{M}_m \tag{2.18b}$$

$$\mathrm{M}_n^* + h\nu_{\text{probe}} \longrightarrow \mathrm{M}_n^+ + \mathrm{e}^- . \tag{2.18c}$$

As a first approximation the obtained ion signal is proportional to the number of excited aggregates. In this approach the ion signal which is recorded as a function of the time delay Δt between the pump and probe pulses gives information directly about the population of the excited state at

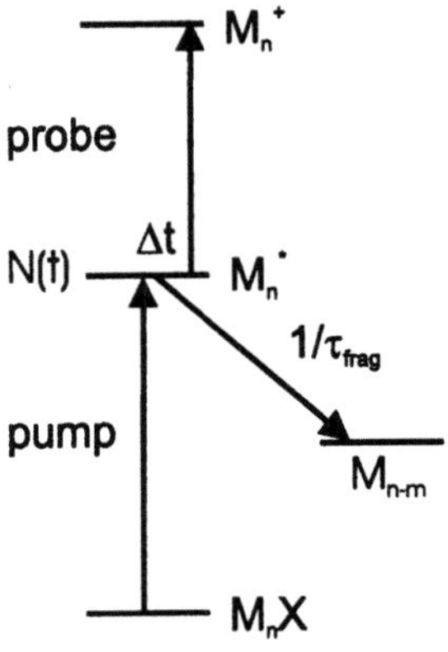

Fig. 2.35. Model of photoinduced fragmentation. The schematic energy-level system represents the excitation (pump), ionization (probe), and fragmentation of the excited state M_n^*. A pump pulse is used to transfer the initial population N_0 from the ground state $M_n X$ to the excited state M_n^*. The population $N(t)$ can be probed by ionization with a second pulse as a function of time t. For this the probe pulse is successively delayed in time (Δt). Owing to fragmentation with a probability $1/\tau_{\text{frag}}$, the population of M^* decreases exponentially and a fragment M_{n-m} is formed.

the well-defined time $t = \Delta t$. Hence, the real-time evolution of the ion signal reflects the transient change of the excited state population.

With these assumptions, the temporal evolution of the excited state can be explained in the following way. The pump pulse will generate an initial population N_0 in the excited state within its pulse width. Owing to fragmentation with a probability $1/\tau_{\text{frag}}$, this population decreases with time t. The decay is characterized by the lifetime τ_{frag} of the excited ensemble.

Hence, the temporal change of the excited population at the time t is proportional to the number $n(t)$ of excited aggregates at that time t multiplied by the probability $1/\tau_{\text{frag}}$. Owing to fragmentation, the excited systems break into two or even several daughter systems, reducing the initially generated population N_0. This photoinduced dissociation process is reflected by a decreasing ion signal.

As a resumé, the temporal change of the population density $n(t)$ of the excited state can be described by the following rate equation:

$$\frac{\mathrm{d}}{\mathrm{d}t}n(t) = N_0\delta(t) - \frac{1}{\tau}n(t). \tag{2.19}$$

Assuming δ-shaped excitation laser pulses, which generate the initial population N_0 of excited aggregates, for $t > 0$ the solution of this simple differential equation is given by

$$n(t) = N_0 \exp(-\frac{t}{\tau}). \tag{2.20}$$

To take account of the temporal width of the laser pulses the real-time spectra have to be compared with the convolution function

$$f(t) = \ell(t) * n(t), \tag{2.21}$$

where $\ell(t)$ is the overall system response of the measuring system. The overall system response is represented by the cross correlation of the laser pulses.

This simple energy-level model has been successfully tested by analyzing real-time TPI experiments with picosecond time resolution carried out in

1991 on the D state of Na_3 [306] and is used as a basis for the analysis of the measurements discussed in Chap. 4. In particular, it is applied to the real-time investigations of the predissociated Na_3 C state in Sect. 4.1.

Extended Energy-Level Model. The real-time spectra – performed with femtosecond time resolution – of the Na_3 system, as well as of the larger sodium and potassium clusters (see Sects. 4.2, 4.3 and 4.4), reveal a nonexponential decay which cannot be explained within the simple energy-level model introduced above. It seems reasonable that this different behavior is caused by clusters in the beam which are larger than those of interest. Therefore, the simple model has to be slightly extended. This extended energy-level model has the following features.

- Direct fragmentation of the clusters of interest excited by the pump pulse. This fragmentation process is identical to that of the simple decay model and is called Type I. Its time constant is given by τ_1.
- Increasing population due to fragmentation of larger clusters into the recorded mass channel with a decay constant τ_m, followed by a re-fragmentation of these clusters with time constant τ_2 (Type II fragmentation)

Figure 2.36 sketches this extended fragmentation model.

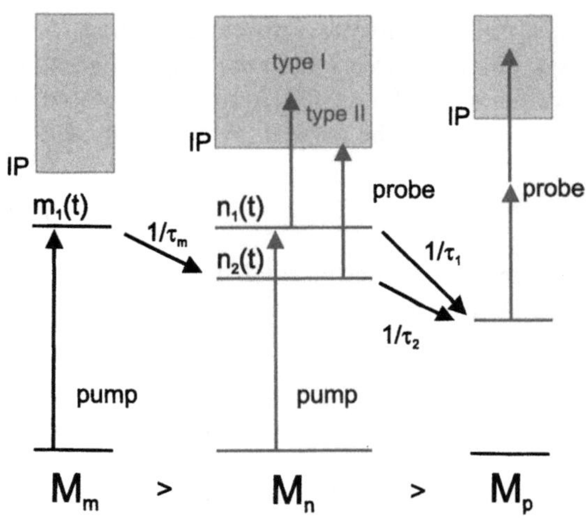

Fig. 2.36. Extended fragmentation model (taken from [307]). For explanation see the text in this section

From now on the number of clusters involved in the Type I and Type II dissociation process will be called $n_1(t)$ and $n_2(t)$, respectively. The number of larger clusters, which are excited by the pump pulse and dissociate into the recorded mass channel, is characterized by a single simplifying function $m_1(t)$. Hence, this extended fragmentation model leads to the following system of coupled differential equations:

$$\frac{\mathrm{d}}{\mathrm{d}t} n_1(t) = N_1 \delta(t) \frac{1}{\tau_1} - n_1(t), \tag{2.22a}$$

$$\frac{\mathrm{d}}{\mathrm{d}t} n_2(t) = \frac{1}{\tau_\mathrm{m}} m_1(t) - \frac{1}{\tau_2} n_2(t), \tag{2.22b}$$

$$\frac{\mathrm{d}}{\mathrm{d}t} m_1(t) = M_1 \delta(t) - \frac{1}{\tau_\mathrm{m}} m_1(t). \tag{2.22c}$$

The temporal evolution of Type I and Type II dissociation is fully described – for a δ-shaped pump pulse – by (2.2.2a) and (2.2.2b), respectively. Here N_1 is the initial population prepared by the pump pulse. Equation (2.2.2c) summarizes the overall dynamic behavior of all larger clusters involved in the Type II process. M_1 is the initial population prepared by the pump pulse.

As indicated in (2.23a) the measured ion signal is given by the sum of the Type I (2.23b) and the Type II (2.23c) contributions determined by the solution of the differential equation system

$$n(t) = n_1(t) + n_2(t) \tag{2.23a}$$

$$= N_1 \exp(-t/\tau_1) \tag{2.23b}$$

$$+ \frac{M_1 \tau_2}{\tau_\mathrm{m} - \tau_2} \left[\exp(-\frac{t}{\tau_\mathrm{m}}) - \exp(-\frac{t}{\tau_2}) \right]. \tag{2.23c}$$

Neglecting the Type II fragmentation, a single exponential decay (2.23a) is obtained, corresponding to the simple fragmentation model. Owing to the temporal width of the laser pulses, the measured ion signal $n_\mathrm{ion}(t)$ can again be well approximated by a convolution $f_\mathrm{ext}(t)$ of $n(t)$ with the cross correlation of the pump and probe pulses, called the 'overall system response' to the laser pulses $\ell(t)$:

$$f_\mathrm{ext}(t) = n(t) * \ell(t). \tag{2.24}$$

The function $f_\mathrm{ext}(t)$ is fitted to the experimental real-time decay curves $n_\mathrm{ion}(t)$ by means of a least-squares-fit routine, based on the 'downhill simplex method' [308].

2.2.3 Theoretical Description of Ultrafast Structural Relaxation

The excitation of a molecule or cluster by an ultrashort laser pulse – as is shown for silver aggregates in Chap. 5 – induces time-dependent changes of their electronic and atomic structure. These changes can be nicely studied by real-time experiments inducing charge reversal processes (see the 'NeNePo' scheme in Fig. 1.1). Very recently, several rather different theoretical approaches have been published or presented at conferences dealing with the relaxation mechanisms induced by ultrafast excitation processes. A brief summary of these theories is given here.

First, Bennemann, Garcia, and coworkers [134, 135] described the ultrafast structural response of the optically excited Ag_3 molecule by combining a microscopic electronic theory with molecular dynamics simulations in the

Born–Oppenheimer approximation. In particular, they analyzed the time evolution of the ionization potential and the dependence of the dynamics on the initial temperature of the Ag_3 molecule. The simulations were performed by use of the Verlet algorithm in its velocity form.

Shortly later, Bonačić-Koutecký and Warken [309] applied an ab initio quantum mechanical treatment of stationary and dynamical properties of bound vibrational systems to the relaxation dynamics of Ag_5 induced by an electron photodetachment process. With their approach they are able to calculate stationary and vibrational properties of vibrational systems with up to 30(!) nonseparable degrees of freedom. The method is based on the representation of the Watson Hamiltonian [310] in spaces spanned by harmonic-oscillator eigenfunctions. Energies and eigenvectors of the Hamiltonian are obtained by a vibrational configuration interaction method. The time propagation of the vibrational states can be regarded as the key issue of their studies. For this, two different approaches are employed: a direct integration of the Schrödinger equation and the residue generation method [311]. With this scheme the short-time vibrational relaxation following the photodetachment of an electron from Ag_5^- clusters was analyzed.

Together with Jortner, the group of Bonačić-Koutecký [136] also treat theoretically the charge reversal process observed for Ag_3. This study is based on the time-dependent ab initio quantum chemical density matrix method in the Wigner representation. The essential point in their calculations and simulation of the charge reversal processes is the correct treatment of the optical excitation during the finite duration of the detachment and ionization pulse. This determines the distribution and decay of the phase space density and as a consequence the real-time evolution of the observed ion signal. The optical excitation is described by a classical Franck–Condon transition using the method of Wigner distributions. The Franck–Condon transition is modulated with a time-dependent phase, which contains the energy spacing of the involved electronic states determined by classical molecular dynamics.

The simulation of Bonačić-Koutecký starts with the generation of an anionic ensemble (1000 molecules) with initial values of space and velocity coordinates. For this, classical molecular dynamics on the ground state PES of the anion is performed. Followed by the Franck–Condon transition (pump pulse), the ensemble is propagated on the PES of the neutral. Next the time-dependent vertical energy gaps between the neutral and anion PESs (vertical detachment energy), as well as between the cation and neutral PESs (vertical ionization energy), are extracted. From these data the real-time photoion spectrum is calculated via generation of ensemble mean values.

V. Engel and coworkers [312] also investigated the time-resolved charge reversal process in molecules and applied their theory to the organometallic molecule ironcarbonyl (FeCO). A direct grid propagation of nuclear wave packets was used. Compared with [136], Engel states that the main purpose of their treatment is especially to take into account the ionization continua produced within the charge reversal process studied.

3. Wave Packet Propagation

In recent years, the wave packet approach has proved to be a powerful tool to describe and analyze the real-time dynamics of molecular motion [2, 3]. Many of the theoretical investigations in this field [7, 8, 299, 313] depend on the work of Heller, who formulated a basic description in 1981 [284]. Parallel to this, magnificent experimental studies in the field of laser femtochemistry, pioneered by Zewail, were successfully performed [20].

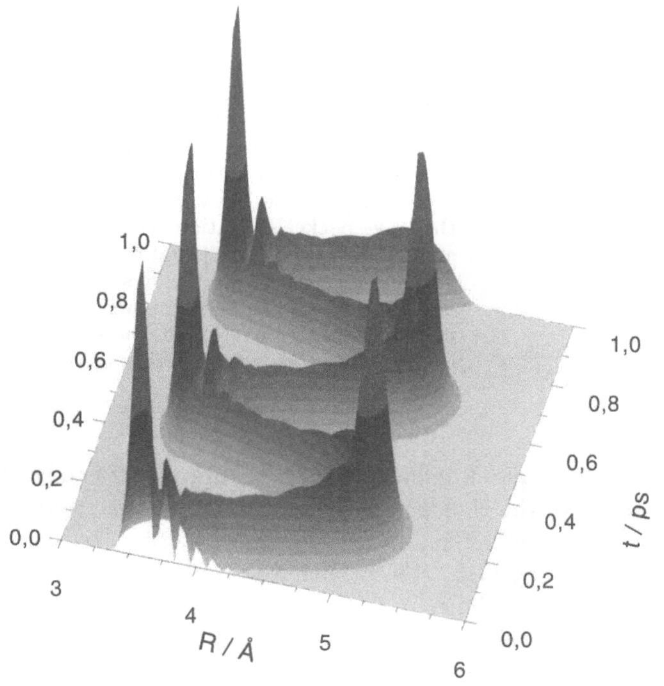

Fig. 3.1. Wave packet propagation on the lowest-lying excited electronic state (A) of NaK (taken from [314]). An oscillation with a period ~ 450 fs is seen. More details are given in Sect. 3.1.6

The central idea of this ultrafast spectroscopy of molecules is the preparation of molecular wave packets followed by the detection of their motion in real time. A wave packet is defined as a coherent superposition of several energy eigenstates [6, 7, 9]. Hence, to prepare a wave packet, laser pulses with a spectral width ΔE_ℓ which is broad compared to the level spacing ΔE_m of the molecular eigenenergies have to be applied. In the time domain, this implies that relative to the period $T = 2\pi\hbar/\Delta E_\mathrm{m}$ of the molecular motion considered, the duration of the preparing pulse has to be short. Vibrational periods of small molecules are of the order of 10^{-14} to 10^{-12}s [10]. Lasers generating pulses in the sub-100 fs regime made feasible the real-time spectroscopy of these systems. Figure 3.1 nicely presents the propagation of a wave packet on the NaK A state potential-energy surface. The wave packet moves back and forth on the surface as the $\sim$ 450 fs oscillation shows.

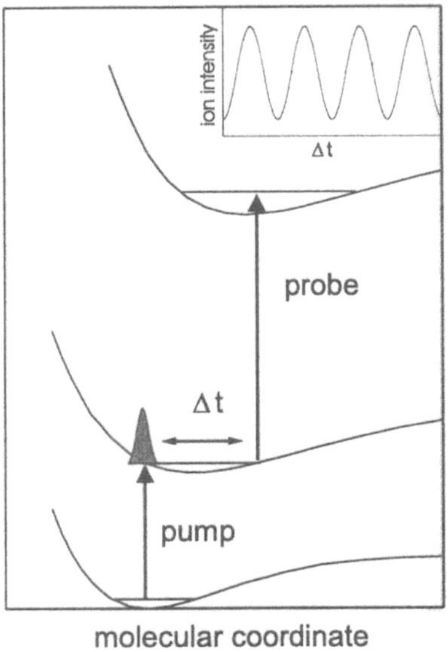

Fig. 3.2. Principle of real-time TPI spectroscopy. A wave packet is prepared in an excited state of a neutral molecule by a pump pulse. Since in general the transition probability to the ion state is a function of the wave packet's location on the potential-energy surface, the propagation of the wave packet can be probed by a second, time-delayed pulse. The inset shows a possible real-time spectrum $I(t)$, i.e. the intensity of the detected ions as a function of the delay time Δt

The different probe techniques implemented in real-time spectroscopy can be distinguished in terms of the choice of the final state. The first, and still most common, probe technique applied in femtosecond experiments is time-resolved electronic spectroscopy [315, 316]. In such experiments, the prepared excited state is projected onto a higher-lying electronic state, and the fluorescence from this final state is collected. Another method involves measuring the excited state absorption of the probe beam [317]. Alternatively, the dynamics can be projected back onto the ground state of the neutral (stimulated emission pumping) [318] or onto the ground state of the ion by means of photoionization [18]. Figure 3.2 exemplarily illustrates this technique for the case of TPI spectroscopy. A potential advantage of this techniques is that the final state is often well characterized by independent methods such as high resolution spectroscopy or ab initio computation.

The choice of the ground state of the ion as the final state $| \phi_f \rangle$ may have several conceptual and practical advantages. First, the detection of charged particles is extremely sensitive. Second, the detection of the ions provides mass information, and third, the ionization is always an allowed process. Any molecular state can be ionized, whereas the electronic spectroscopy relies on the existence of optically allowed transitions. Besides this, no Rabi oscillation between bound states, which can interfere with wave packet measurements, can occur in ionization. Further information can be obtained by analyzing the photoelectron (e.g. as to its kinetic energy), analogous to dispersed fluorescence methods.

To observe the pure principles of wave packet propagation up to now, one has had to concentrate the investigations mainly on small, i.e. di- and triatomic, molecular systems. Therefore, this chapter is divided into two parts. First, in Sect. 3.1, the ultrafast dynamics of different alkali dimers is investigated in greater detail. Besides some basics, several features, for example energy dependence, isotopic effects, controlled molecular dynamics, and revivals of wave packets, are discussed. The use of the spectrogram technique nicely visualizes the different phenomena.

In the second part of this chapter (Sect. 3.2), different wave packet propagation phenomena in excited alkali trimers are discussed. The time-resolved pseudorotation of the sodium trimer is presented in Sect. 3.2.2. Last but not least, applying laser pulses of the same wavelength but of different pulse width enables a mode-selective preparation of the trimer, hence controlling its dynamics (Sect. 3.2.4). Wave packet propagation on a repulsive PES (Sect. 3.2.5), studied on the potassium trimer, leads to the phenomena of ultrafast photodissociation, which then is the topic of the subsequent chapter.

3.1 Femtosecond Wave Packet Propagation Phenomena in Excited Alkali Dimers

In this section real-time spectroscopy of various alkali dimers is presented. The section is divided into several parts, each presenting a different mechanism which influences the wave packet dynamics in a particular way. The potassium dimer acts as a kind of workhorse. For moderate pump laser intensities, fascinating wave packet propagation phenomena can be revealed. In Sect. 3.1.1 a few essentials of the wave packet propagation are described using K_2 as a model system. Further exciting phenomena can be studied on the dimers. Applying different probe wavelengths not only changes the pump&probe cycles but also can influence the detected wave packet propagation, as is demonstrated in Sect. 3.1.2. Perturbations, due for example to a crossing electronic state, influence and therefore can drastically change the laser-induced wave packet dynamics of a dimer. Two examples are given here. Exciting two isotopes of K_2 with pulses of the same wavelength show totally different behavior (Sect. 3.1.3). The wavelength dependent spin-orbit coupling of the A and b states of the sodium dimer strongly governs the induced wave packet dynamics. This is presented in Sect. 3.1.4. Changing the intensity of the pump laser pulse as well can substantially influence and even be used to control molecular dynamics, as is shown in Sect. 3.1.5. Owing to the absence of a center of symmetry, heteronuclear dimers are also interesting candidates for the study of wave packet propagation phenomena. The NaK molecule, treated in Sect. 3.1.6, reveals an amazing interplay of total and fractional revivals of the wave packet dynamics.

3.1.1 K_2 A State at Moderate Laser Intensities

In this section a brief introduction is given, presenting a few general features of a dimer's real-time MPI spectrum. Special details are then discussed in the following sections.

The first excited electronic state, the $A\,^1\Sigma_u^+$, of the potassium dimer is an ideal candidate to act here as the model system. The A state of K_2, like the other alkali dimer A states, is more strongly bound than the ground state $X\,^1\Sigma_g^+$ owing to its partial ionic character. From the viewpoint of classical spectroscopic experimental techniques, the A state of $^{39,39}K_2$ has been studied by laser-induced fluorescence (LIF), optical–optical double resonance, and Fourier transform spectroscopy [319–323]. Ro-vibrational levels could be identified and the spectroscopic constants were calculated by a Dunham fit. An RKR (Rydberg–Klein–Rees) analysis was used to deduce the PES of K_2. A strong spin–orbit coupling between the A state and the superimposed $b^3\Pi_u$ (Fig. 3.3) was observed around the vibrational quantum state with $v = 12$, but this is the first topic of Sect. 3.1.3.

Figure 3.3 depicts the excitation scheme of the one-color three-photon ionization (3PI) process investigated here. Before light interaction, the dimer is – owing to its generation by adiabatic expansion – in the vibrational state $v' = 0$ of the electronic ground state $X\,^1\Sigma_g^+$. Taking into account the known spectroscopic data [322–324] on the ground and A states as well as the Franck–Condon principle, the pump pulse prepares a wave packet on the A state PES close to the inner turning point. This vibrational wave packet $\psi(x,t)$ can be described by

$$\psi(x,t) = \sum_n a_n \mid v_n\rangle \exp\left(-\mathrm{i}\frac{E_n t}{\hbar}\right) \tag{3.1}$$

where $\mid v_n\rangle$ is the set of vibrational eigenstates E_n of the A state and the a_n are constant coefficients, which contain both the amplitudes and initial phases of the excited vibrational states $\mid v_n\rangle$. Now the Hamiltonian of the

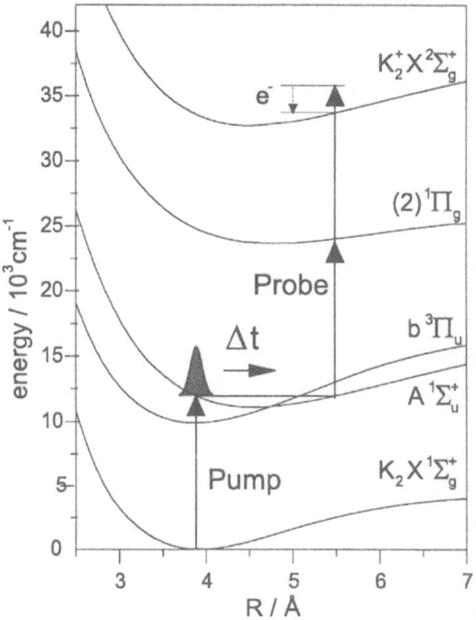

Fig. 3.3. Excitation scheme for a one-color real-time (3PI) experiment on K_2 (taken from [262]). A wave packet is prepared in the $A\,^1\Sigma_u^+$ by an ultrashort pump pulse. Its propagation on this PES is probed by a two-photon ionization process. The intermediate state $(2)\,^1\Pi_g$ acts as a Franck–Condon window. Excess energy is given to the electron e^-. The propagation of the wave packet is perturbed by the $b\,^3\Pi_u$ state (see Sect. 3.1.3). The data for the PES are taken from [322, 324–326]

excited system can do its work, i.e. the generated wave packet propagates on the A state PES. After a certain delay time Δt this wave packet is probed using two photons trying to ionize the dimer ($K^+\ X\,^2\Sigma_g^+$). The electron (e^-), separated during the ionization process, takes the excess energy as kinetic energy.

A resonant intermediate state ($(2)\,^1\Pi_g$) acts as a spatially small Franck–Condon window. As is seen during the analysis of the real-time spectrum, this state with parity 'gerade' acts under these special excitation conditions as a kind of filter, since only for a single internuclear separation R (Condon point) of the dimer can the transition take place from the A state to the ion state. This fact enables the explicit mapping of the wave packet's motion, i.e. the signal is expected to come and go periodically as the wave packet travels through this detection window.

The ion signal $I(t)$ is detected as a function of the time delay between pump and probe pulses. As a result the wave packet propagation of interest can be monitored in real time as an oscillation of the ion signal I. The period of the ion signal is related to the classical frequency ω_{cl} of the molecular vibration. For example, if the Condon point is at a turning point, one would expect the signal to oscillate at frequency ω_{cl}. The signal is

$$I(t) = \mid P(t) \mid^2, \tag{3.2}$$

where

$$P(t) = \langle \phi_f \mid \boldsymbol{\mu} \cdot \boldsymbol{E} \mid \psi(x,t) \rangle = \sum_n b_n \exp(-\mathrm{i}\frac{\mathrm{E_n t}}{\hbar}). \tag{3.3}$$

Here $\langle \phi_f \mid$ represents the final ionic state, $\boldsymbol{\mu}$ the transition dipole moment, $\boldsymbol{E}$ the electric field vector of the probe laser pulse, and

$$b_n = a_n \langle \phi_f \mid \boldsymbol{\mu} \cdot \boldsymbol{E} \mid v_n \rangle. \tag{3.4}$$

Therefore

$$I(t) = \sum_{n,m} b_n b_m \cos\left[\frac{(E_n - E_m)t}{\hbar}\right] \tag{3.5}$$

is expected. This ion signal is composed of beat frequencies between all pairs of energy levels (E_n, E_m) that make up the wave packet. The lowest frequencies correspond to beats between adjacent vibrational levels $n = m \pm 1$, with a frequency of approximately ω_{cl}.[1] There are as well higher frequencies corresponding to beats between further seperated energy levels. In general, frequencies of approximately $q\omega_{\mathrm{cl}}$ are due to beats between levels n and $n-q$, where q is an integer.

The resulting real-time spectrum of a one-color three-photon ionization (3PI) is shown in Fig. 3.4 over a range of about 200 ps. The excitation wavelength was $\lambda = 833.7$ nm (12 040.81 cm^{-1}). The quadrupole mass spectrometer was aligned to the maximum ion yield of the $^{39,39}K_2$ isotope of the potassium dimer and the step width of the delay unit was 50 fs.

The temporal evolution of the ion signal reveals a clear oscillation. Its period is about 500 fs. The maximum oscillation amplitude is close to the zero of time and it decreases within 5 ps, but growing again rather rapidly. The long-time behavior shows a repetition of this behavior similar to beat structures, as known for example in acoustics when two frequencies close to each other are superimposed. As can be seen in Fig. 3.5, a closer look at the zero-of-time gives further information.

- First, the ion signal is symmetric about the zero of time. This fact is easy to understand since in a one-color pump&probe experiment, at $\Delta t = 0$ the pump and probe pulses change their roles, i.e. the pump pulse becomes the probe and vice versa.
- Second, an oscillation period $T_{\mathrm{A}}^{(833)} \approx 501$ fs is estimated as the mean of the real-time spectrum. Comparing this with spectral data for K_2, $T_{\mathrm{A}}^{(833)}$ can be assigned to the vibrational period of a wave packet excited to the first excited electronic state, the A state of the dimer.
- Third, the first maximum is not found at $\Delta t = 0$, but at $T_{\Phi,\mathrm{A}} = 250$ fs. This gives important information about the transition pathway during the induced three-photon ionization process. The pump pulse prepares the wave packet at the inner turning point of the A state. Then the wave

[1] Due to the anharmonicity of the A state potential energy surface the level spacing between different vibrational eigenstates decreases with growing energy.

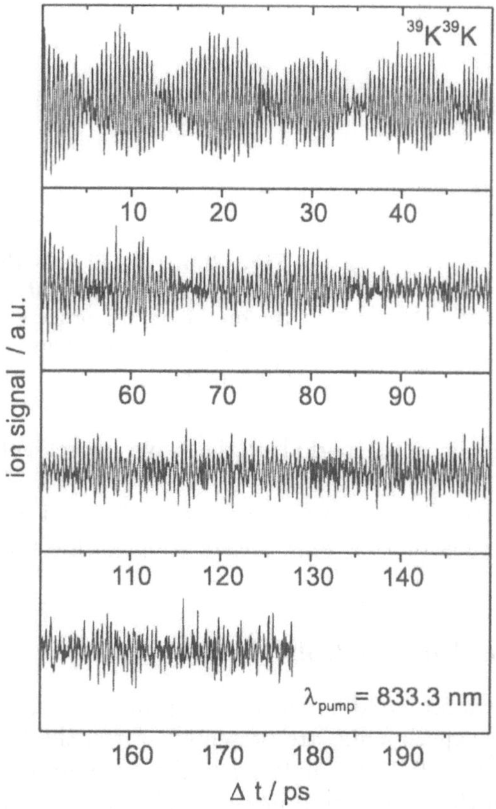

Fig. 3.4. Real-time spectrum $I(\Delta t)$ of the one-color 3PI process of $^{39,39}K_2$ (taken from [43])

packet propagates and after 250 fs the wave packet has reached the PES's outer turning point. As the real-time spectrum proves, here a maximum of the ion signal is found. Then it takes a further 500 fs, that is, a full round trip (back and forth) of the wave packet on the PES. Hence, under these excitation conditions it can be concluded that the ionization step always takes place at the outer turning point of the PES.

To get further information from these real-time data, Fourier analysis is an appropriate tool. In particular, the origin of the beat structure can be analyzed (see Sect. 3.1.3).

Figure 3.6 shows the wavelength dependence of the real-time spectra. The oscillation period decreases slightly with growing excitation wavelength λ. Since the accessory frequency can be identified with the spacing between the excited vibrational levels, this result demonstrates that real-time spectroscopy allows an easy and direct scanning of the PES and its anharmonicity. Owing to the anharmonicity of the PES, for lower excitation wavelengths the spacing between the vibrational levels is larger, hence the oscillation fre-

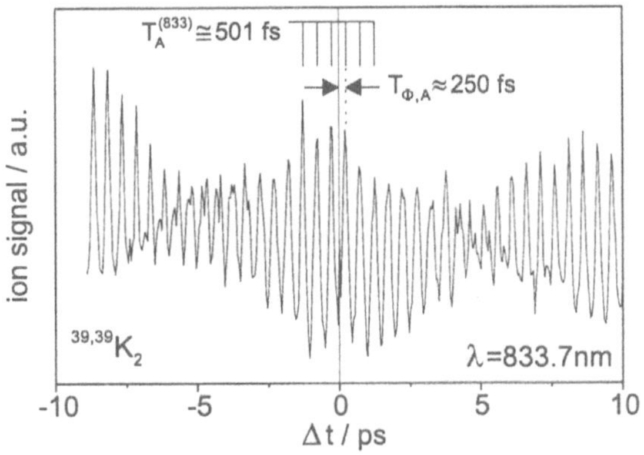

Fig. 3.5. Detail of the real-time spectrum $I(\Delta t)$ of the one-color 3PI process of $^{39,39}K_2$ at delay times Δt close to zero

quency is bigger and as a consequence the period of the wave packet gets smaller.

With this, the first general view of the real-time spectra is finished and now a detailed analysis of several fascinating wave packet propagation phenomena follows.

3.1.2 K_2 A State for Different Pump&Probe Cycles

In this section, the real-time spectra for two different pump&probe cycles, i.e. two-color and one-color real-time experiments, are presented. First, the two essentially different real-time spectra $I(t)$ are analyzed in the time domain, followed by a detailed analysis in the frequency domain ($I(\omega)$). Introducing the spectrogram technique ($I(\Delta t, \omega)$) enables detailed insight into the investigated wave packet dynamics. In particular, it visualizes directly several total and fractional revivals of induced vibrational wave packets. By comparing the spectrograms of the different pump&probe cycles, one can easily assign the different ionization pathways of the laser-induced processes. This nicely demonstrates the different excitation mechanisms in the two experiments.

Most of the real-time experiments on alkali dimers performed so far have used – as introduced above – three-photon ionization (see also Sects. 3.1.3–3.1.6). However, applying an ultrastable femtosecond laser system combined with a long-time stable molecular beam allows also a straightforward approach by two-photon ionization. In this case the observed wave packet propagation in the $^{39,39}K_2$ A $^1\Sigma_u^+$ state is not at all superimposed by any influence of an intermediate probe step.

Time Domain. The results of the one-color and two-color experiments are compared in Fig. 3.7. The Figures 3.7 a, b show the raw pump&probe data. The output voltage U_{ion} of the secondary electron multiplier is plotted as a function of the delay time Δt between the pump and probe pulses. In both cases (two-color experiment, Fig. 3.7 a and one-color experiment, Fig. 3.7 b), an oscillatory time-dependent variation superimposed on a time-independent

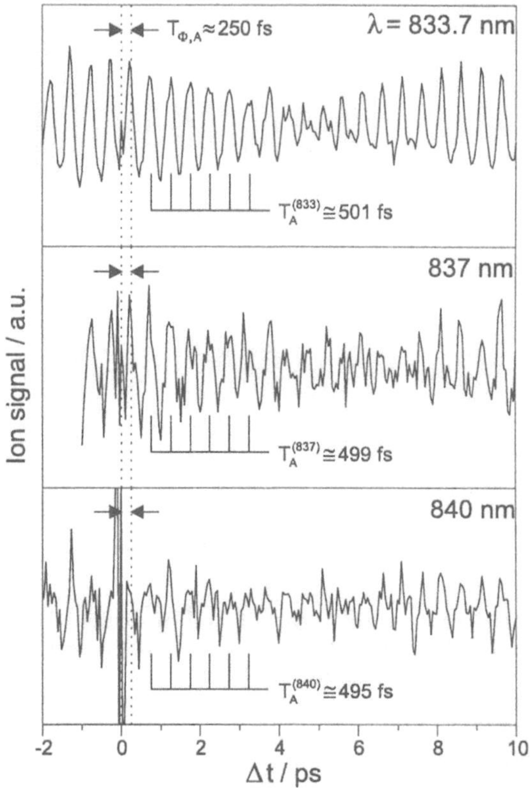

Fig. 3.6. Real-time spectrum $I(\Delta t)$ of the one-color 3PI process of $^{39}K_2$ for different excitation wavelengths (taken from [262])

background signal is found. Since in these panels the raw data are presented, one can directly estimate the contribution of the time-dependent ion signal in comparison to the total ion signal. To characterize the time-dependent contribution, it is useful to look at the ratio κ of the amplitude of the time-dependent modulation $\Delta U_{\rm dyn}$ in the ion signal to the voltage $U_{\rm n}$, which defines the zero line of the observed oscillatory pattern.

For the two-color and one-color experiments the values of κ are 0.08 and 0.47 respectively. The rather large value of κ for the one-color experiment can be explained by the multiphoton ionization process in the probe step [42]. A wave packet prepared by the pump pulse on the $^{39,39}K_2$ $A\,^1\Sigma_u^+$ potential-energy surface can be detected via a two-photon ionization step induced by the probe pulse. As shown in Sect. 3.1.1, this two-photon ionization process is strongly enhanced by the intermediate $(2)\,^1\Pi_{\rm u}$ state while the wave packet is located at the outer turning point of the A state PES. Therefore, the wave packet propagation in the A state can be detected as a rather pronounced oscillation [42] of the ion signal as is nicely seen in Fig. 3.7 b. The observed main period, $T_{\rm A,1} \cong 500$ fs, can be directly assigned to the classical vibrational period of a K_2 molecule excited to its electronic A state. Besides this, the

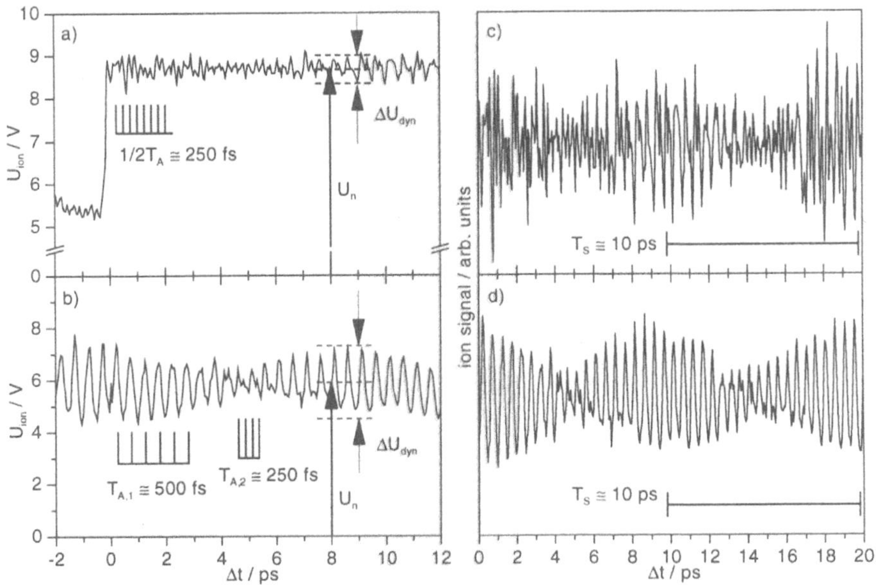

Fig. 3.7. Real-time spectra $I(\Delta t)$ from two-color (**a, c**) and one-color (**b, d**) experiments on $^{39,39}K_2$. In (**a**) and (**b**) the raw data are presented, while in (**c**) and (**d**) the normalized data sets are plotted (taken from [52])

period $T_{A,2} \cong 250$ fs is observable in the region around the delay time $T_{\text{frev},1} \cong 5$ ps. This oscillation time is nearly half the classical period. The reason for this period will be discussed in the next subsection. These two oscillation patterns are superimposed on a beat structure with a period $T_S \cong 10$ ps. Its pronounced dynamics are nicely observable in Fig. 3.7 d, where the normalized data are plotted. Note that in the one-color experiment the fast oscillation with $T_{A,2} \cong 250$ fs appears as well-centered around the delay time of $T_{\text{frev},2} \cong 15$ ps. The beat structure with maxima at $T_{\text{rev},1} \cong 10$ ps and $T_{\text{rev},2} \cong 20$ ps is interpreted in Sect. 3.1.3 as a sequence of strongly shifted revivals of the wave packet oscillation in the A state [43]. The real-time spectrum of the one-color experiment is symmetric about the zero-of-time. This is due to the exchange of the roles of the identical pump and probe pulses at the temporal origin.

For the two-color experiment (Fig. 3.7 a) the situation is totally different. In this case the pump and probe pulses have different wavelengths and, hence, different PESs are excited by the pump pulse. Therefore, at the zero of time it is switched between two totally different excitation mechanisms. For $\Delta t > 0$ the Franck–Condon factor A $\longleftarrow$ X from the ground state $X\,^1\Sigma_g^+$ (vibrational level $v'' = 0$) to the A state is optimal for a laser of wavelength 834 nm [323]. Since ionization from the A state by a laser of wavelength 417 nm is energetically possible all along the propagation path of the prepared wave packet, an intense time-independent background can be expected (see Fig. 3.7 a). For negative delay times, however, no PES can be directly reached by the laser of $\lambda = 417$ nm [325]. Hence, the probe pulse of 834 nm cannot resonantly ionize molecules from an excited electronic state. As a consequence, the background signal for $\Delta t < 0$ is much lower than for positive delay times.

Superimposed on the strong background for $\Delta t > 0$, a weak oscillation with period $\frac{1}{2}T_{A,1} \cong 250$ fs is detected. An additional oscillatory structure with a 500 fs period is also observable, especially around $\Delta t = 10$ ps. The correlation of the period observed here with the oscillation of the wave packet on the A state PES is discussed below. For the two-color experiment a beat structure with approximately 10 ps period is also visible in the real-time spectrum (more clearly visible in Fig. 3.7 c).

Frequency Domain. Now, the frequency spectra of the one-color and two-color experiments are compared. To get comparable Fourier spectra, the normalized data presented in Figs. 3.7 c, d for delay times 0 ps $< \Delta t <$ 30 ps are used for the calculation of the Fourier transform.

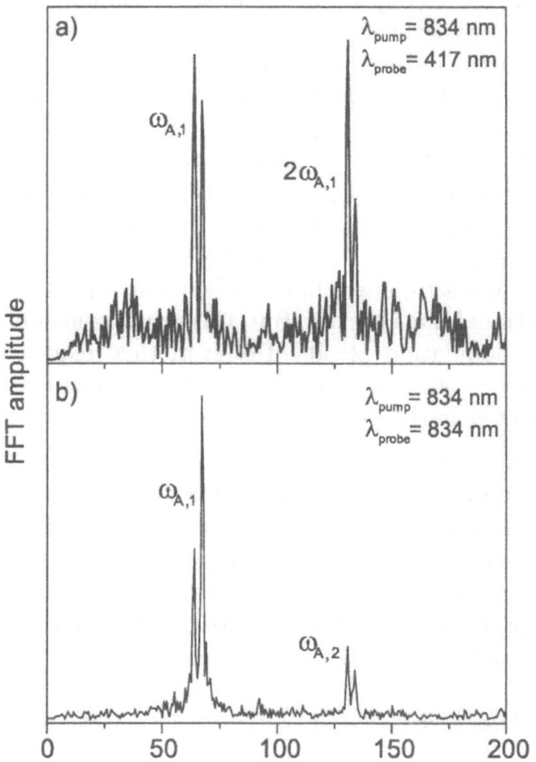

Fig. 3.8. Fourier spectra $I(\omega)$ of the real-time spectra of the $^{39,39}K_2$ two-color (**a**) and one-color experiment (**b**) (taken from [52])

The Fourier spectra are presented in Figs. 3.8 a (two-color experiment) and 3.8 b (one-color experiment). Frequency components appear in both spectra at nearly the same position. In the frequency spectrum of the two-color experiment (Fig. 3.8 a) two frequency bands at $\omega_{A,1} \cong 66\,\mathrm{cm}^{-1}$ and $2\omega_{A,1} \cong 132\,\mathrm{cm}^{-1}$ are detected. The two bands have about the same intensity and both show a fine structure with two main components. As discussed in detail in Sect. 3.1.1, the Fourier spectrum of the one-color experiment (Fig. 3.8 b) also reveals two frequency bands in this spectral range. Here, the main frequency band $\omega_{A,1} \cong 66\,\mathrm{cm}^{-1}$ and a minor band $\omega_{A,2} \cong 132\,\mathrm{cm}^{-1}$ are observed. The intensity of $\omega_{A,2}$ amounts to approximately 25% of the main frequency's intensity.

The frequency band $\omega_{A,1}$ can be assigned to the classical frequency of the wave packet in the $\mathrm{A}\,^1\Sigma_u^+$ state, where each component of the band is correlated with the energy spacings of the perturbed vibrational energy levels [43]. The period corresponding to the main frequency band is $T_A \cong 500\,\mathrm{fs}$. The double structure of $\omega_{A,1}$ corresponds to the beat oscillation observed in the real-time spectra. Owing to a coherent excitation of vibrational energy levels (quantum number v') with $\Delta v' = 2$, the frequency $\omega_{A,2}$ appears in the spectrum of the one-color experiment (Fig. 3.8 b). As in the band $\omega_{A,1}$, the effect of perturbation-induced energy-level shifts has similar consequences for the distribution of frequencies in the second band $\omega_{A,2}$.

Figure 3.8 a shows that the wave packet dynamics on the A state's PES can be visualized by the two-color experiment, although the ionization is energetically possible all along the internuclear distance. However, for the two-color experiment the intensity of the $2\omega_{A,1}$ band is nearly equal to that of the $\omega_{A,1}$ band. The frequency $2\omega_{A,1}$ is correlated with the relative strong 250 fs oscillation in the real-time spectrum (Fig. 3.7 a). By comparison with the one-color experiment, one might assign each line of this frequency to a coherent excitation of three vibrational energy levels. But, this would contradict the distribution of the excited population density. However, this frequency component can be well understood by closer examination of the corresponding spectrogram $I(\Delta t, \omega)$ presented in the next subsection.

Spectrograms. To obtain deeper insight into the dynamics of the wave packet, it is of great interest to know the temporal evolution of the frequency components, i.e. the time Δt when and with what intensity I a certain frequency ω occurs in the real-time spectra. This information can easily be extracted from spectrograms $I(\Delta t, \omega)$ [51, 262, 327]. For this purpose a procedure is used which first was proposed by Stolow et al. to calculate spectrograms of transient data [51]. A sliding-window Fourier transform

$$I(\omega, \Delta t) = \int_0^\infty I(t) g(\tau - \Delta t) \mathrm{e}^{-\mathrm{i}\omega\tau} \mathrm{d}\tau \tag{3.6}$$

is used, where $I(\tau)$ is the real-time ionic signal at time τ. For the window function g a Gaussian shape

$$g(\tau) = \exp\left(-\frac{\tau^2}{t_0^2}\right) \tag{3.7}$$

is chosen with $t_0 = 1\,\text{ps}$. The Fourier amplitude can be plotted as a function of the delay time Δt and the frequency ω. This enables the direct observation of the time dependence (Δt) of the different frequency (ω) components originating from the propagating wave packet, including their relative contributions I. In particular, the interplay of the frequency groups involved can be seen at a glance. Revivals, total and fractional [54], will be emphasized in the spectrograms. Hence, the introduction of the spectrogram technique for analysis of ultrafast molecular dynamics can be regarded as a highly promising method. Up to now, spectrograms of molecular wave packets have been presented for halogen dimers [51, 327] and for alkali dimers [52, 53, 262]. However, compared to the Fourier analysis as shown in Fig. 3.8 there is quite a large loss of frequency resolution.

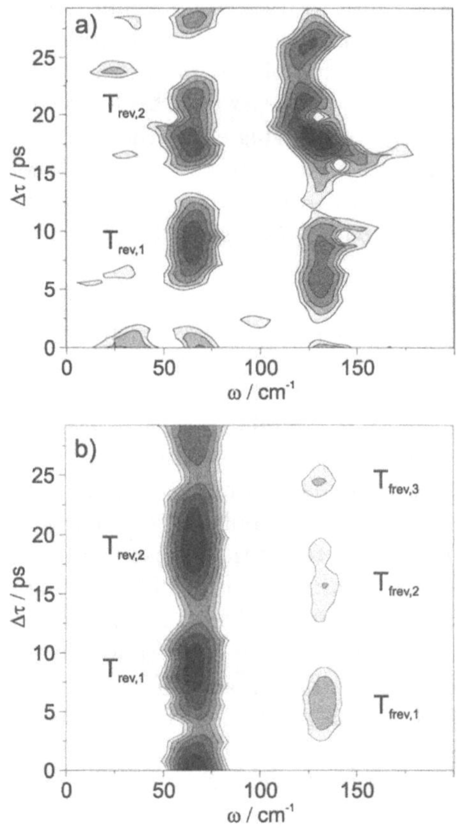

Fig. 3.9. Logarithmic contour plots of the spectrograms $I(\Delta, \omega)$ for the two-color (**a**) and one-color (**b**) experiment in $^{39,39}K_2$ (taken from [52]). The Fourier amplitude increases from white to black

The logarithmic contour plots of the spectrograms are presented in Fig. 3.9a (two-color experiment) and Fig. 3.9b (one-color experiment). The Fourier amplitude is plotted with increasing gray from white to black. The frequency bands already seen in the frequency spectra (Fig. 3.8) can now be examined with their temporal evolution.

For the one-color experiment (Fig. 3.9 b) the main frequency band $\omega_{A,1}$ (corresponding to $T_{A,1} \cong 500\,\text{fs}$) appears as expected every 10 ps. There, at $T_{rev,1} \cong 10\,\text{ps}$ and $T_{rev,2} \cong 20\,\text{ps}$, the wave packet initially prepared by the pump pulse runs into total revivals. As pointed out in Sect. 3.1.3 and [43], the fairly short revival time of 10 ps is caused by the spin–orbit coupling between the $A\,^1\Sigma_u^+$ and $b\,^3\Pi_u$ states. The second band, $\omega_{A,2}$ ($T_{A,2} \cong 250\,\text{fs}$), caused by coherent superposition of vibrational energy levels with $\Delta v' = 2$, is also visible every 10 ps at $T_{frev,1} \cong 5\,\text{ps}$, $T_{frev,2} \cong 15\,\text{ps}$, and $T_{frev,3} \cong 25\,\text{ps}$. A distinct alternation between $\omega_{A,1}$ and $\omega_{A,2}$ is observable. Following the ideas of Averbukh and Perel'man [54], the occurrence of the frequency $\omega_{A,2}$ at half the time of the total revival can be understood as a fractional revival, where the initially prepared wave packet splits into its two components (see fine structure of $\omega_{A,1}$ in Fig. 3.8). In classical terms, these two components of the wave packet represent partial wave packets with a phase difference of π with respect to each other at the times of the fractional revivals. This leads to the oscillation of the partial wave packets of $\omega_{A,1}$ traversing each other. Hence, in the one-color experiment the ionization step, enhanced by the Franck–Condon window built up by the $(2)^1\Pi_g$ state, happens two times inside the classical period $T_{A,1} \cong 500\,\text{fs}$ of the A state wave packet (see excitation scheme in Fig. 3.10, process (b)).

In case of the two-color experiment (Fig. 3.9 a), however, the doubled frequency $2\omega_{A,1}$ ($\hat{=}\frac{1}{2}T_{A,1}$) is of significant intensity at the same time as $\omega_{A,1}$ ($T_{A,1}$) is present, i.e. when the components of the initially prepared wave packet go through a revival. Therefore, we deduce the frequency $2\omega_{A,1}$ observed in the two-color experiment cannot be correlated with a fractional revival. This frequency can be interpreted in the following way. In the pump&probe scheme for the two-color experiment, no mechanism is present to enhance an ionization by a photon of the probe pulse at any particular internuclear distance. The potential-energy curves of the A state and of the ground state of the ion have nearly equal shapes (see Fig. 3.10). Therefore, one can assume a weak dependence of the strength of the ionization step as a function of the internuclear distance. Since the classical dwell time of a wave packet is maximum at the inner and outer turning points of the PES, the transition probability between two PESs of similar shape is highest at these points. Hence, the ionization step is slightly favored at both turning points of the PES of the A state (process (a) in Fig. 3.10). Hence, the doubled frequency of the wave packet propagation in the A state PES is present in the real-time spectrum each time the wave packet undergoes a total revival. This measuring technique, therefore, allows to estimate roughly the dependence of the transition dipole moment $\boldsymbol{\mu}$ on the internuclear distance R, which is mostly set constant in theortical simulations of wave packet propagation. Owing to a rather small variation of $\boldsymbol{\mu}$ the ions can be produced, at least weakly, at all internuclear distances. This manifests in a rather large background signal, as is seen in Fig. 3.7 a.

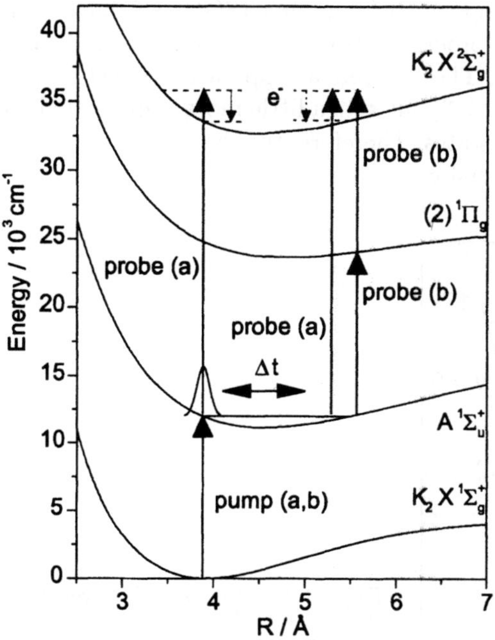

Fig. 3.10. Excitation scheme of $^{39,39}K_2$ for the two-color experiment (a) and one-color experiment (b) (taken from [52]). In both cases a wave packet is prepared on the A state's PES by the pump pulse. In the two-color experiment the subsequent ionization by one photon is slightly favored at both the inner and at the outer turning points of the PES. In the one-color experiment the ionization is strongly enhanced by the $(2)^1\Pi_g$ state acting as a Franck–Condon window. Potential-energy curves are based on data given in [324, 325]

This discussion makes clear that the spectrogram technique enables us to characterize the varied wave packet dynamics for the two different pump&probe cycles. The discussion nicely demonstrates the value of spectrograms for the analysis of wave packet phenomena as total and fractional revivals. Moreover, the spectrograms allow the identification of ionization pathways in pump&probe experiments. Hence, spectrograms are excellent tools for an improved analysis of wave packet propagation phenomena, compared to real-time and Fourier spectra only.

3.1.3 Spin–Orbit Coupled Electronic States: the Isotopes $^{39,39}K_2$ and $^{39,41}K_2$

In this section special interest is focused on the intersection of the intersystem crossing of the K_2 $A\,^1\Sigma_u^+$ state with the $b\,^3\Pi_u$ state. Applying one-color pump&probe spectroscopy with a wavelength of 833.7 nm, the wave packet dynamics of $^{39,39}K_2$ around $v = 12$ is studied. Isotope-selective detection allows us to compare the results with the heavier isotopic dimer $^{39,41}K_2$. Quantum dynamical simulations are used to simulate the results and to discuss the mechanism of the perturbation by a $^3\Pi_u$ state. It is shown that the temporal evolution of the wave packet dynamics is strongly influenced by an intersystem crossing (ISC) process.

Theoretical Considerations. In the present studies of the molecular dynamics of the K_2 $A\,^1\Sigma_u^+$ state, spin–orbit coupling (SOC) effects become important. They enable the ISC process between the two crossing PESs of $A\,^1\Sigma_u^+$ and $b\,^3\Pi_u$ (see Fig. 3.11). The theoretical approach applied is quite similar to that described in Sect. 2.2.2, but gets a little more complicated owing to the coupling of the A and b states. Therefore, the theoretical equipment is summarized briefly for this special case.

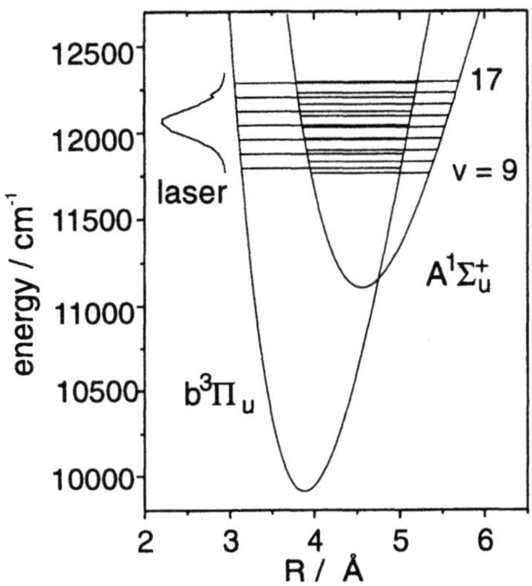

Fig. 3.11. Perturbation region of the K_2 $A\,^1\Sigma_u^+$ state with the $b\,^3\Pi_u$ state. Potential-energy curves and vibrational levels [320] are compared with the spectral position of the excitation laser (taken from [43])

During the pump&probe process a wave packet – similar to that in Sect. 3.1.3 – is prepared in the A state. According to the electronic state selection rules, the b state is dark during femtosecond pulse excitation. From the A state the system can evolve to the b state via SOC. Photoionizing out of the b well is negligible, because the resonant intermediate state $(2)^1\Pi_g$ [42] is again a singlet state. The ISC process indicates the limits of the Born–Oppenheimer approximation. Two basic approaches can be used to describe these electronic transitions, i.e. the adiabatic and diabatic representations [122]. In the present case the SOC between the A and the b state is rather weak ($20\,\mathrm{cm}^{-1}$) and the diabatic representation can be used. This formalism is ideal for use in the matrix formalism, which has already been utilized to simulate the pump&probe ionization process in the absence of ISC effects [42]. Basically, the original PESs, obtained without SOC, are used as electronic basis functions. The molecular Hamiltonian H is expressed as

$$H(Q,r) = T_{\mathrm{nuc}}(Q) + H_{\mathrm{el}}(r,Q), \tag{3.8}$$

with

$$H_{\text{el}}(r,Q) = H_{\text{nr}}(r,Q) + H_{\text{so}}(r,Q), \tag{3.9}$$

where

$$T_{\text{nuc}} = \frac{P^2}{2m} \tag{3.10}$$

is the nuclear kinetic energy operator. The electronic Hamiltonian H_{el} consists of the nonrelativistic electrostatic Hamiltonian H_{nr} and the SOC operator H_{so}. The arguments Q and r represent the nuclear and electronic coordinates.

The first-order spin–orbit interaction splits the $\text{b}\,^3\Pi_\text{u}$ state into its m_j components $\pm 2, \pm 1, 0^\pm$ and, mainly, causes a small energy shift of the coupling $\text{b}\,^3\Pi_\text{u}$ component with respect to the A state. Therefore, the $\text{b}\,^3\Pi_\text{u}$ state is prediagonalized, including first-order spin–orbit effects [320] only. According to the selection rules of H_{so}, only the $m_\text{j} = 0^+$ component of $\text{b}\,^3\Pi_\text{u}$ couples to the A state by second-order interaction, and this is included explicitly in the subsequent theoretical quantum dynamical treatment. The resulting vibronic functions are linear combinations of A and b vibronic functions $\phi_{\text{el,A}}\Phi_\text{A}$ and $\phi_{\text{el,b}}\Phi_\text{b}$, where ϕ_{el} are the electronic and Φ are the nuclear eigenfunctions of the A and b states. The mixing coefficients are proportional to

$$\langle \phi_{\text{el,A}}\Phi_\text{A} \mid H_{\text{so}} \mid \phi_{\text{el,b}}\Phi_\text{b}\rangle \,/\, \left| E_{\phi_{\text{el,A}}\Phi_\text{A}} - E_{\phi_{\text{el,b}}\Phi_\text{b}} \right| . \tag{3.11}$$

The influence of the rotational motion ($T_{\text{rot}} \approx 18\,\text{ps}$, rotational revival $T_{\text{rot,rev}} > 500\,\text{ps}$, estimated from [323]) can be neglected in the simulation, because the investigations concentrate on the short-time dynamics of the ISC process immediately after pump pulse excitation. The time-dependent evolution of the wave packets is evaluated by solving a set of coupled time-dependent Schrödinger equations

$$\text{i}\hbar\frac{\partial}{\partial t}\boldsymbol{\Psi} = \begin{pmatrix} H_{\text{XX}} & H_{\text{XA}} & 0 & 0 & 0 \\ H_{\text{AX}} & H_{\text{AA}} & H_{\text{Ab}} & H_{\text{A(2)}} & 0 \\ 0 & H_{\text{bA}} & H_{\text{bb}} & 0 & 0 \\ 0 & H_{\text{(2)A}} & 0 & H_{\text{(2)(2)}} & H_{\text{(2)I}(E_k)} \\ 0 & 0 & 0 & H_{\text{I}(E_k)\text{(2)}} & H_{\text{I(E}_\text{k})\text{I(E}_\text{k})} \end{pmatrix} \cdot \boldsymbol{\Psi}, \tag{3.12}$$

with

$$\boldsymbol{\Psi} = \begin{pmatrix} \Psi_{\text{X}\,^1\Sigma_\text{g}^+} \\ \Psi_{\text{A}\,^1\Sigma_\text{u}^+} \\ \Psi_{\text{b}\,^3\Pi_\text{u}} \\ \Psi_{(2)\,^1\Pi_\text{g}} \\ \Psi_{\text{I},k}(E_k) \end{pmatrix} \tag{3.13}$$

and the initial conditions

$$\Psi_\text{X}(Q, t=0) = \Phi_\text{X}(Q, v=0). \tag{3.14}$$

The nuclear wave functions Ψ_i are the projections of the total wave functions on the electronic PES involved. They depend on the nuclear coordinate Q and on the time t. Φ_X is the initial vibrational state $v = 0$ of the ground state. The matrix elements of the Hamiltonian (3.12), describing the neutral molecular states, are given by

$$H_{ii} = T_{\text{nuc}} + V_i , \tag{3.15}$$

where V_i represent the PES:

$$V_i = \langle \phi_{\text{el}} \mid H_{\text{nr}}(Q) + H_{\text{so}}(Q) \mid \phi_{\text{el}} \rangle_{\text{r}} . \tag{3.16}$$

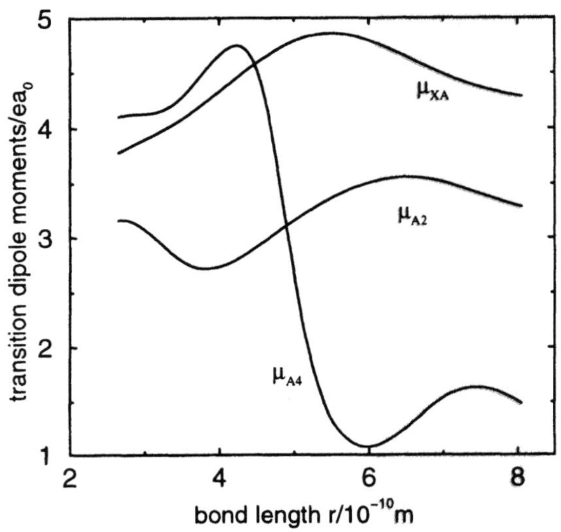

Fig. 3.12. Calculated electronic transition dipole moments μ_{ij} of the $X\,^1\Sigma_g^+ \rightarrow A\,^1\Sigma_u^+$, $A\,^1\Sigma_u^+ \rightarrow 4\,^1\Sigma_g^+$, and of the $A\,^1\Sigma_u^+ \rightarrow 2\,^1\Pi_g^+$ transitions (taken from [328])

The time-dependent off-diagonal elements describe the interaction with the laser field in the dipole approximation and they are given by

$$H_{ij}(t) = -\mu_{ij} E(t) \tag{3.17}$$

with the dipole transition moments μ_{ij} (see Fig. 3.12). The electromagnetic field $E(t)$ is expressed by

$$E(t) = E_{\text{pump}}(t) + E_{\text{probe}}(t + \Delta t) \tag{3.18}$$

with

$$E(t) = E_0 \cos(\omega t) s(t) . \tag{3.19}$$

E_0 is the amplitude of the electromagnetic field, ω the laser frequency, and $s(t)$ a shape function. Gaussian functions are chosen as adequate shape functions for both the pump and the probe pulses. For the time-independent off-diagonal elements $\langle \mathrm{b}\,^3\Pi_\mathrm{u} \mid H_\mathrm{so} \mid \mathrm{A}\,^1\Sigma_\mathrm{u}^+ \rangle$ a constant value of $18.4\,\mathrm{cm}^{-1}$ is taken from [320].

Owing to the ejection of the electron by the probe laser pulse, a continuum is superimposed on the ion ground state. The interval of the total continuum is $[0, 0.38\,\mathrm{eV}]$. Different methods of discretizing the electronic continua have been successfully employed in theoretical studies of ultrafast ionization processes [303–305]. Here, the continuum is simulated by discretizing the corresponding energy range by a sufficient number N of electronic states [303]. The matrix elements representing the continuous part of the Hamiltonian, are given by

$$H_{\mathrm{I}kk} = \langle \phi_{\mathrm{I},k} \mid H_\mathrm{I} \mid \phi_{\mathrm{I},k} \rangle \tag{3.20}$$

and

$$H_{\mathrm{I}ki} = \langle \phi_{\mathrm{I},k} \mid \mu_\mathrm{I}(E_k) E(t) \mid \phi_i \rangle\,, \tag{3.21}$$

with the electronic basis functions of the continuum $\phi_{\mathrm{I},k}$ and the electronic wave functions of the neutral state ϕ_i.

Equation (3.12) is then solved without further approximation and within a model including all multiphoton processes. The solution of (3.8) is obtained by the second-order differencing fast Fourier transform (SOD-FFT) method [287, 288, 299]. The relevant PESs for the multiphoton ionization process considered here are the X, A, b, $(2)^1\Pi_\mathrm{g}$, and ion states. The PESs and the transition dipole moments are obtained from ab-initio data (Sect. 3.1.5 and [328]).

Perturbed Wave Packet Propagation. The influence of a perturbation on the propagation of the wave packet is now presented in the time as well as the frequency domain. Finally, the mechanism of the build-up dynamics and the time dependence of the induced electronic population dynamics are discussed.

Time Domain. The pump&probe spectra for $^{39,39}\mathrm{K}_2$ and $^{39,41}\mathrm{K}_2$ were recorded for delay times between -5 ps and more than 180 ps. The temporal evolution of the ion signal's intensity for both isotopes is shown in Fig. 3.13 for delay times between 0 and 200 ps. In both pump&probe spectra a fine oscillatory structure with an oscillation period $T_\mathrm{A} \approx 500$ fs – the full 2π oscillation time of the wave packet in the A state – is present over the whole range. The first maximum of the oscillation appears at a delay time of 250 fs, half (π) of the A state period, when the ionization step takes place at the outer turning point [42]. This fine oscillation is superimposed on a long-time evolution which reveals totally different features for the two isotopes: in the transient of the $^{39,39}\mathrm{K}_2$ signal (Fig. 3.13 a) a beat structure with a period $T_\mathrm{BS} \approx 10$ ps dominates in the temporal region shown in the figure. A double structure

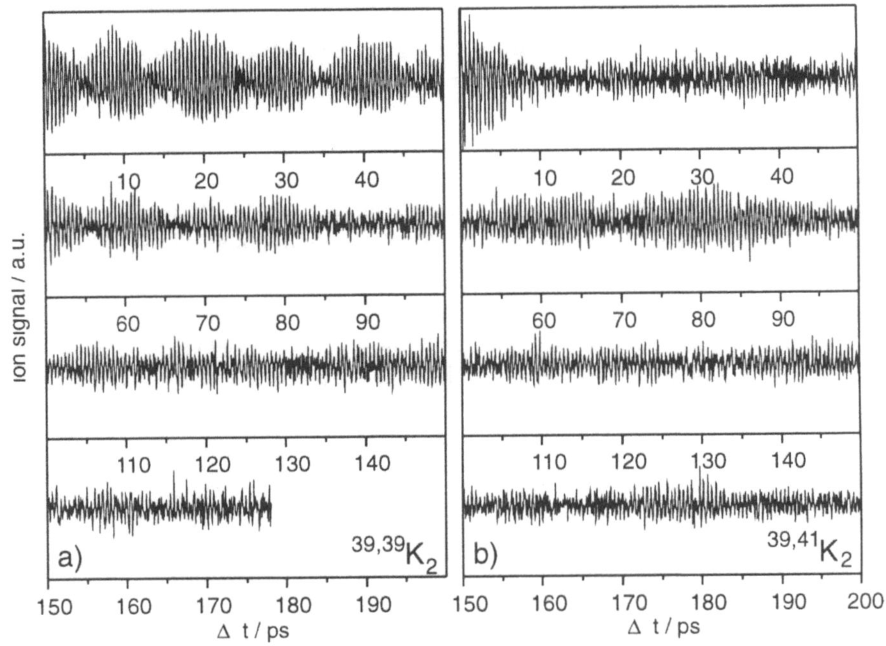

Fig. 3.13. Real-time 3PI signal for the two K_2 isotopes **(a)** $^{39,39}K_2$ and **(b)** $^{39,41}K_2$ (taken from [43])

of beat oscillation maxima appears at $T_{BD,1} \approx 10$ ps and at $T_{BD,2} \approx 60$ ps. For $^{39,41}K_2$ (Fig. 3.13 b) the long-time structure includes different features: a regular dephasing and some fractional revivals [54] are observed. The main revivals appear at 38 ps, 60 ps, and 82 ps. The spectrograms (Fig. 3.14) nicely present the different observed wave packet propagation behaviors of the two isotopes.

Quantum dynamical calculations of the pump&probe spectra for the two isotopes were performed for delay times up to 40 ps. A comparison of the experimental and theoretical ionization signals as a function of the delay time is presented in Fig. 3.15. In agreement with the experimental data, the short-time dynamics of the theoretical signal show the 500 fs oscillation period of the wave packet prepared in the $A\,^1\Sigma_u^+$ state (centered around $v = 11$) and the long time dynamics reflect the totally different beat structures of the two isotopes. However, the oscillation periods of the pronounced regular beat structure of the isotope $^{39,39}K_2$ (Fig. 3.15 a) and of the weak, irregular beat structure of the isotope $^{39,41}K_2$ (Fig. 3.15 b) are somewhat shorter for the theoretical signal. In the case of $^{39,39}K_2$ a period of $T_{BS} \approx 7.5$ ps and in the case of $^{39,41}K_2$ a period of about 20 ps is found. The double structure observed in the $^{39,39}K_2$ signal during one beat oscillation period around the delay times $T_{BD,1}$ and $T_{BD,2}$ (Fig. 3.13 a) appears even more strongly in the simulation (see Fig. 3.15).

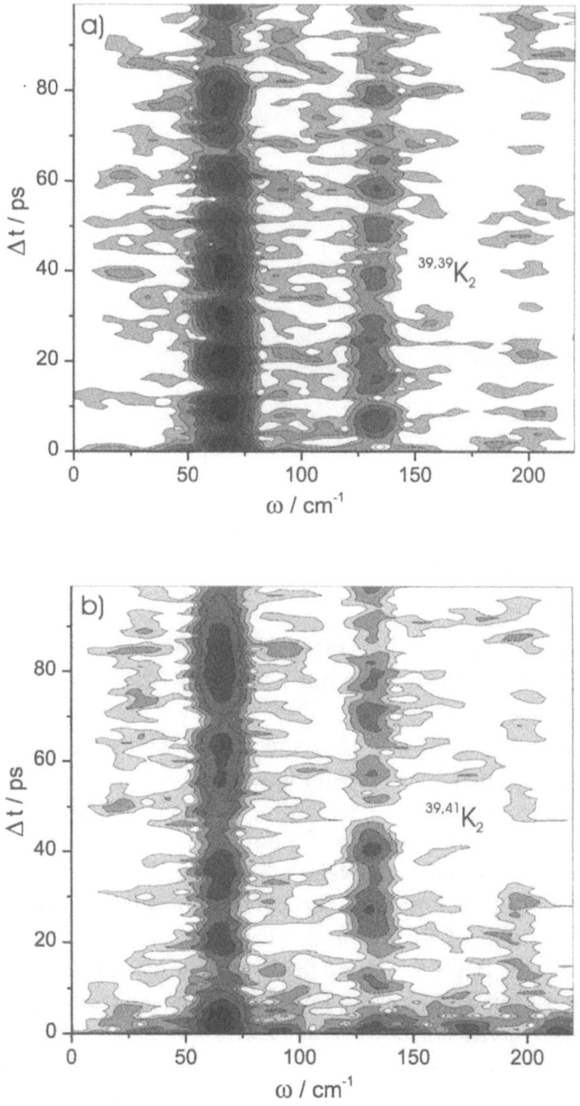

Fig. 3.14. Logarithmic contour plots of the spectrograms $I(\Delta, \omega)$ of the two isotopes **(a)** $^{39,39}K_2$ and **(b)** $^{39,41}K_2$ (taken from [262]). The Fourier amplitude increases from white to black

Frequency Domain. The frequency components involved in the transient spectra are extracted by a Fourier analysis of the normalized real-time data (see Fig. 3.16). Both Fourier spectra are dominated by a group of frequencies around $\omega_0^{(1)} \approx 65\,\mathrm{cm}^{-1}$. Two additional frequency groups with lower amplitudes appear at $\omega_0^{(2)} \approx 130\,\mathrm{cm}^{-1}$ and at $\omega_0^{(3)} \approx 195\,\mathrm{cm}^{-1}$. An additional peak is observed at $\omega_X \approx 90\,\mathrm{cm}^{-1}$, where the relative intensity is slightly larger in the case of the lighter isotope (Fig. 3.16a).

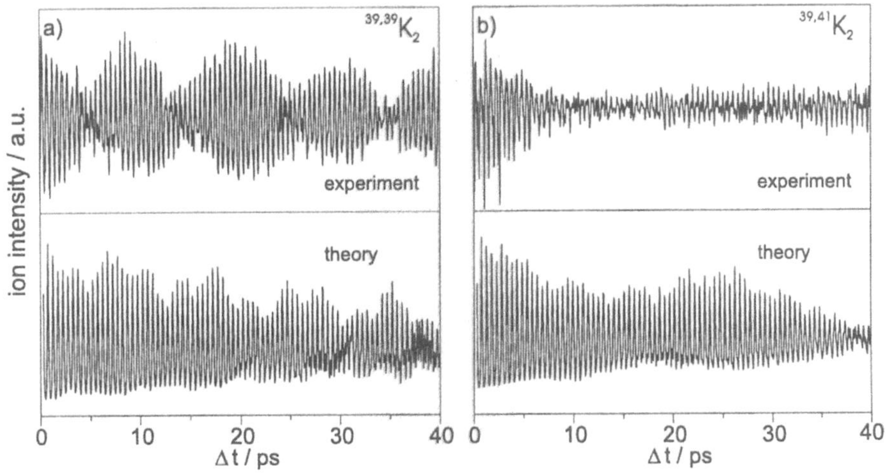

Fig. 3.15. Comparison of experimental real-time spectra and quantum dynamical real-time simulation for (**a**) $^{39,39}K_2$ and (**b**) $^{39,41}K_2$. The top panels contain experimental data and the bottom panels theoretical data (taken from [43])

The frequency group around $\omega_0^{(1)}$ is illustrated in the insets of Fig. 3.16 for the case of the isotopes of K_2 studied. The inset of Fig. 3.16(a) shows the Fourier components in the real-time spectrum of $^{39,39}K_2$. Two main frequencies at $63.8\,cm^{-1}$ and $67.2\,cm^{-1}$ dominate this spectrum. The component at $67.2\,cm^{-1}$ seems to be broadened by a component at $66.8\,cm^{-1}$. Further frequency components can be observed at $64.7\,cm^{-1}$, $65.2\,cm^{-1}$, $65.6\,cm^{-1}$, and $67.9\,cm^{-1}$. Hence, the wave packet consists of a non-monotonic frequency distribution of the contributing vibrational levels, which manifests itself as a spectral hole. Instead of the frequency values, the corresponding vibrational level pairs are given. These pairs of vibrational levels are found by introducing an energetic shift to the RKR levels of [323] and by comparing the resulting energy spacings between neighboring levels with the Fourier analysis data (see Table 3.1). Introducing shifts for $v = 12$ and $v = 13$ of $1.2\,cm^{-1}$ and $2.1\,cm^{-1}$ respectively results in a good agreement. The two dominant frequencies observed in the real-time spectrum can be clearly proved to be responsible for the beat oscillation period of $T_{BS} \approx 10$ ps: overlaying the frequencies $\omega_{13,14} = 63.8\,cm^{-1}$ and $\omega_{12,13} = 67.2\,cm^{-1}$ leads to a period of $T_{BS} = 9.8$ ps. For the isotope $^{39,41}K_2$, the spectrum around $65\,cm^{-1}$ is presented in the inset of Fig. 3.16 b. Five distinct components in the frequency group at $64.3\,cm^{-1}$, $64.7\,cm^{-1}$, $65.5\,cm^{-1}$, $65.9\,cm^{-1}$, and $66.4\,cm^{-1}$ are observed. Here, the numbering is not included owing to missing spectroscopic data for this isotope.

The Fourier spectra were also calculated for the theoretical data. Since the simulations were performed up to 40 ps only, two main frequencies $\omega_{calc\,1,2}$ are resolved (see Table 3.1). The resulting frequencies $\omega_{calc\,1} = 66.8\,cm^{-1}$ and

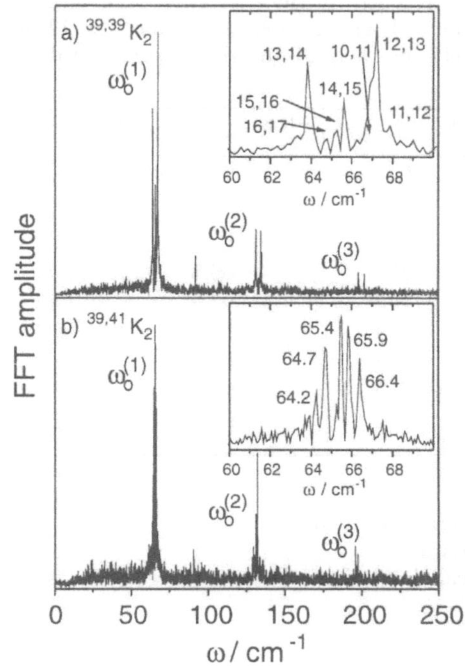

Fig. 3.16. Fourier spectra calculated from the normalized pump&probe data, (**a**) for $^{39,39}K_2$ and (**b**) for $^{39,41}K_2$ (taken from [43]). The insets show details of the Fourier spectra between 60 and $70\,cm^{-1}$. The inset of part (**a**) contains the vibrational level pairs belonging to the frequency components, while in the inset of (**b**) the frequency values are given

$\omega_{calc\,2} = 62.5\,cm^{-1}$ lead to the period $T_{BS} = 7.8\,ps$. The difference between the experimental and theoretical findings in the ion signal originates from the deviation of the ab initio PES from the observable potential (compare with [323]).

Experimental and theoretical Fourier spectra reveal the effect of the SOC on the $b\,^3\Pi_u$ state. In the region of the pump laser pulse the perturbation is most effective in the isotope $^{39,39}K_2$. Two vibrational levels of the $b\,^3\Pi_u$ state ($v = 23, 24$) are close [320], nearly energetically degenerate, to the two vibrational levels of the $A\,^1\Sigma_u^+$ state ($v = 12, 13$) and induce the shift of these perturbed A state vibronic levels (as seen in the inset of Fig. 3.16 a). For $^{39,41}K_2$ the situation is different. The perturbing levels in the excitation range of the pump pulse are not energetically close and all the vibrational levels contributing to the wave packet are perturbed, but only by a small amount (see Fig. 3.16 b). The regular pattern of an unperturbed spectrum (see e.g. [320]) is conserved.

Mechanism. Finally, the build-up dynamics and the time dependence of the induced electronic population dynamics are investigated. Subsequently we consider whether, in addition to the vibrational dynamics, the time-resolved electronic population dynamics are detectable in the ion-signal.

The temporal evolution of the radiationless transition is illustrated by a sequence of representative wave packets $\Psi_{A\,^1\Sigma_u^+}(Q,t)$ and $\Psi_{b\,^3\Pi_u}(Q,t)$ resulting from the solution of the coupled Schrödinger equation (3.12) in the time

Table 3.1. Vibrational RKR level pairs reached by the excitation laser and corresponding energy spacings without and with energy level shifts for $v = 12$ and 13 of $1.2\,\mathrm{cm}^{-1}$ and $2.1\,\mathrm{cm}^{-1}$ respectively, in comparison with experimental and theoretical data for $^{39,39}K_2$

$v, v+1$	$\omega_{RKR}/\mathrm{cm}^{-1}$, [a]	$\omega_{shift}/\mathrm{cm}^{-1}$, [b]	$\omega_{FFT}/\mathrm{cm}^{-1}$, [c]	$\omega_{calc}/\mathrm{cm}^{-1}$, [d]
9,10	67.35	[e]	[f]	[f]
10,11	67.03	[e]	66.8	[f]
11,12	66.71	67.91	67.9	[f]
12,13	66.40	67.30	67.2	66.8
13,14	66.08	63.98	63.8	62.5
14,15	65.76	[e]	65.6	[f]
15,16	65.44	[e]	65.2	[f]
16,17	65.11	[e]	64.7	[f]
17,18	64.47	[e]	[f]	[f]

[a] Vibrational level spacings of [323].
[b] Level spacings with introduced shifts for $v = 12, 13$.
[c] Fourier components of pump&probe data.
[d] Main Fourier components of theoretical simulation.
[e] Not changed.
[f] Not observed.

domain $0.1\,\mathrm{ps} \leq t \leq 3.0\,\mathrm{ps}$ (Fig. 3.17). From $\Psi_{A\,^1\Sigma_u^+}(Q,t)$ and $\Psi_{b\,^3\Pi_u}(Q,t)$, the total population P of the electronic states $A\,^1\Sigma_u^+$ and $b\,^3\Pi_u$ is derived:

$$P(A\,^1\Sigma_u^+, t) = \int_{Q_l}^{Q_u} \left| \Psi_{A\,^1\Sigma_u^+}(Q,t)\Psi_{A\,^1\Sigma_u^+}(Q,t+\Delta t) \right| \mathrm{d}Q \tag{3.22}$$

and

$$P(b\,^3\Pi_u, t) = \int_{Q_l}^{Q_u} \left| \Psi_{b\,^3\Pi_u}(Q,t)\Psi_{b\,^3\Pi_u}(Q,t+\Delta t) \right| \mathrm{d}Q, \tag{3.23}$$

with $Q_l = 5.0a_0$ and $Q_u = 15.0a_0$ the lower and upper limits of the spatial grid. The total populations of both states are presented for two different time periods, up to 3 ps in Fig. 3.18 and up to (20 ps) in Fig. 3.19.

First, the ISC process for the isotope $^{39,39}K_2$ will be discussed. The results presented in Figs. 3.17 and 3.18 point to the details of how the ISC process is built up. Figure 3.17 shows selected snapshots of representative wavepackets during the first oscillation period of the electronic populations $P(A\,^1\Sigma_u^+, t)$ and $P(b\,^3\Pi_u, t)$ shown in Fig. 3.18.

The wave packet is created by the pump laser at the inner turning point of the A state (first snapshot in Fig. 3.17). Energetically the wave packet is located about 0.1 eV above the electronic curve crossing. The ISC occurs for the first time at about 250 fs (second snapshot in Fig. 3.17), when the center

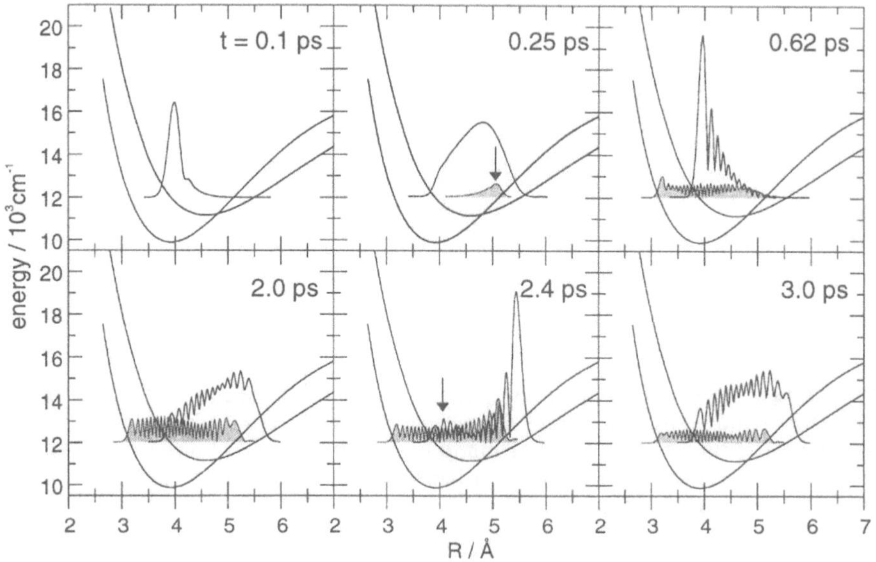

Fig. 3.17. Snapshots of the ISC process induced by SOC between the $A\,^1\Sigma_u^+$ state and the $b\,^3\Pi_u$ state of $^{39,39}K_2$ during and after pump pulse excitation for selected delay times (taken from [43])

of the wave packet $\Psi_{A\,^1\Sigma_u^+}(Q,t)$ approaches the outer turning point of the b state (at $Q = 9.5a_0$). It is to be noted that the transition $A\,^1\Sigma_u^+ \longrightarrow b\,^3\Pi_u$ takes place at the outer turning point of the b state and not at the location of the curve crossing at $Q = 8.9a_0$, contrary to other findings [58].

As soon as some fraction of $\Psi_{b^3\Pi_u}(Q,t)$ has been created, the wave packet begins to propagate according to its own time-dependent phase relation (see Fig. 3.17). When the wave packet $\Psi_{b^3\Pi_u}(Q,t)$ passes the inner turning point of the A state ($Q = 8.22a_0$), some fraction of $\Psi_{b^3\Pi_u}(Q,t)$ is transferred back into the A state (see Fig. 3.17 for the delay times 0.62 ps and 2.4 ps). These back-and-forth transitions give rise to the fine oscillations superimposed on the low-frequency beating in the population probabilities $P(A\,^1\Sigma_u^+,t)$ and $P(b\,^3\Pi_u,t)$ (Figs. 3.18 and 3.19). The oscillation period of 224 fs corresponds to the vibrations of a wave packet in the corresponding 'adiabatic' potential, characterized by the inner turning point of the A state and the outer turning point of the b state. Evidently, the vibrational energy transfer in a bound-state system from one electronic state into another occurs with the highest probability at the turning point of the actual receiver state. Possible reasons are of course favorable Franck–Condon factors of the wave packets and the near-zero kinetic energy of the newly prepared wave packet.

Similar findings for coupled bound-state systems are reported for several model studies by Stock and Domcke [55, 56]. For dissipative systems Zewail and his group [329, 315] observed a step function superimposed on

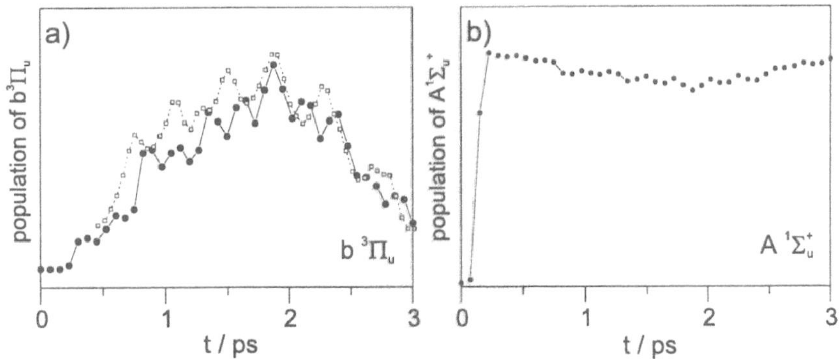

Fig. 3.18. Electronic population in $A\,^1\Sigma_u^+$ and $b\,^3\Pi_u$ states of $^{39,39}K_2$ during and after pump pulse excitation for a time period of 3 ps (taken from [43]). **(a)** Population of the $b\,^3\Pi_u$ state after pump (—) and probe (...) pulse excitation. The probe pulse intensity is reduced by a factor of 10. **(b)** Population in the A state after excitation with the pump pulse

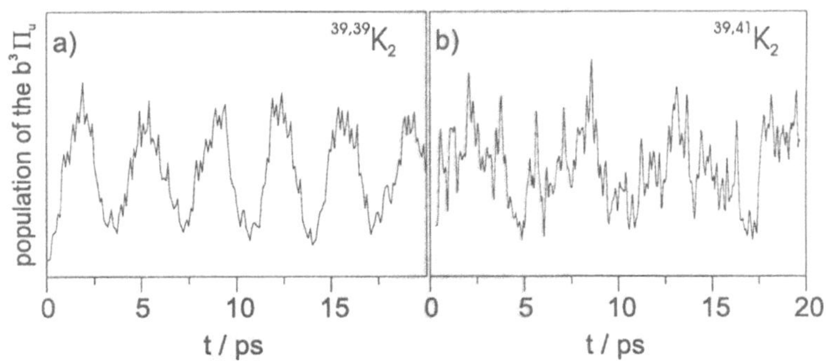

Fig. 3.19. Electronic population in $A\,^1\Sigma_u^+$ and $b\,^3\Pi_u$ states during and after pump pulse excitation for the isotopes **(a)** $^{39,39}K_2$ and **(b)** $^{39,41}K_2$ for a propagation time of 20 ps (taken from [43])

an exponential increase, revealing in this case one optimal location for the non-Born–Oppenheimer transition in one direction only. In $^{39,39}K_2$ the low-frequency oscillation period of the electronic population is proportional to the effective perturbation matrix element between them. The first maximum of the b state population is reached at 2 ps (Fig. 3.18), which means that the $A\,^1\Sigma_u^+ \longrightarrow b\,^3\Pi_u$ ISC occurs within a time-scale of $\tau_{ISC} \approx 1.87$ ps. The relative changes of the amplitude of $P(A\,^1\Sigma_u^+, t)$ are rather small (about 8%). The dotted curve in Fig. 3.18 a presents the population in the b state after probe pulse interaction. Fine oscillations are still recognizable, but evidently the probe laser pulse slightly averages over the fast oscillations.

The period of the low-frequency beating is 3.754 ps (Fig. 3.19), which corresponds to the average energy splitting (about 8 cm^{-1}) of the two closely spaced pairs of A and b vibronic levels. This oscillatory structure remains constant (Fig. 3.19a) and is reflected in the double peak structure superimposed on the 10 ps beat structure in the ion signal (Figs. 3.13 a and 3.15). As pointed out already, the beating period $T_{BS} \approx 10$ ps originates from the shifted vibronic levels in the $A\,^1\Sigma_u^+$ state. For delay times larger than 80 ps, when the beat structure vanishes in the ion signal, the fast oscillations in the $b\,^3\Pi_u$ population also begin to disappear. Both observations are due to the known spreading of the wave packet in the A state [42].

For the less perturbed isotope $^{39,41}K_2$, the perturbation is more delocalized over several vibrational levels. The amount of population and energy transfer is smaller and slower (Fig. 3.19). The resulting time-scale of ISC induced by SOC is $\tau_{ISC} \approx 6$ ps and the electronic population oscillates with a longer period of 7 ps. Moreover, the low-frequency beating loses its pronounced characteristics (Fig. 3.19) and consequently no structure is superimposed on the ion signal.

These combined experimental and theoretical investigations demonstrate that the femtosecond pump&probe technique can be used as a highly sensitive method to study excited vibrational states and their perturbation due to crossing electronic states. It is to be noted that the beat structure in $^{39,39}K_2$ is due to the spectral hole in the frequency distribution of the A state wave packet. It is, however, not related to interferences of wave packets propagating on different electronic PESs as reported for Na_2 at 620 nm [28]. From the theoretical treatment it is possible to observe the build-up of an ISC process which strongly influences the temporal evolution of the 3PI signal in the perturbed case. This analysis reveals that under the given experimental conditions it is possible to detect the low frequency mode of the electronic population dynamics superimposed as a double peak structure on vibrational dynamics in the 3PI signal.

3.1.4 Spin–Orbit Coupled Electronic States of Na_2

In Sect. 3.1.3, wave packet dynamics influenced by spin–orbit coupling of crossing PESs was investigated. It was demonstrated that, although the applied excitation wavelength was identical, the wave packet propagation gave an unambiguous fingerprint of each isotope of a potassium dimer. Now, the influence of the excitation wavelength will be studied for a single isotope. The sodium dimer can be regarded as a model system for this investigation. First, as a basis of the discussion, the real-time spectra obtained by pump&probe experiments for two excitation and ionization wavelengths, 620 nm and 642 nm, are presented. Subsequently, their Fourier spectra are examined and compared with spectroscopic data from earlier experimental work.

Spectroscopic Basics. The sodium dimer's first singlet electronic state $A\,^1\Sigma_u^+$ has been studied by different cw techniques using laser-induced fluorescence, optical–optical double resonance, and Fourier transform spectroscopy [319, 320, 330–332]. Ro-vibrational levels could be numbered and the spectroscopic constants were calculated by a Dunham fit. A Rydberg–Klein–Rees (RKR) analysis was used to deduce the potential-energy curves. Effantin and coworkers [320] observed a strong spin–orbit coupling (SOC) between the $A\,^1\Sigma_u^+$ state and the crossing $b\,^3\Pi_u$ state (intersystem crossing) for nearly every vibronic eigenstate at high rotational quantum number ($J \gg 20$). However, they could not estimate the line shifts for the vibrational level $v = 8$ ($8 < J_{pert} < 14$) because the perturbation here is very close to its origin. Therefore, in their excellent paper [320] the value is only missing for $v = 8$.

The influence of ISC processes on molecular dynamics has been highlighted theoretically for some examples [55–58]. However, until now there has existed only a small amount of experimental information [43] about the effect of SOC on the propagation of wave packets. Na_2 excited to the vibronic level $v = 8$ of the $A\,^1\Sigma_u^+$ state, therefore, is an excellent candidate to investigate this. Comparing the real-time spectra measured close to $v = 8$ and rather far away from the perturbation's origin demonstrates the strong influence of SOC on the wave packet propagation and allows us to determine the resulting shift directly.

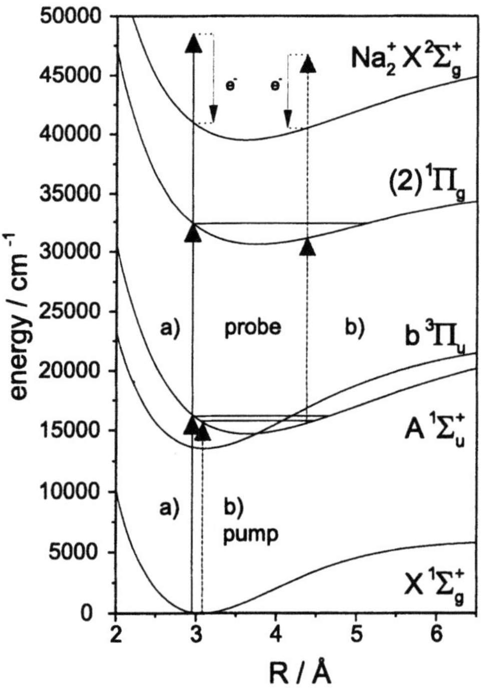

Fig. 3.20. Potential-energy surfaces of Na_2 involved in the three-photon ionization process (taken from [333]). Different transition pathways (**a**) and (**b**), found for excitation with $\lambda_{pump} = 620\,nm$ and $\lambda_{pump} = 642\,nm$ respectively, are indicated

Figure 3.20 presents the potential-energy surfaces of Na_2 involved in the investigated 3PI process. A wave packet is prepared in the A state by the pump pulse. The ionization step needs two photons of the probe pulse. The indicated transition pathways (a) and (b) are discussed later.

Wavelength-Dependent Real-Time Spectra of Na_2. Real-time spectra obtained at 620 nm and 642 nm, are presented in Figs. 3.21 a and b respectively. For 620 nm measurements were continued to times up to 60 ps, steps of 20 fs, while for the longer wavelength a scan up to 50 ps was performed.

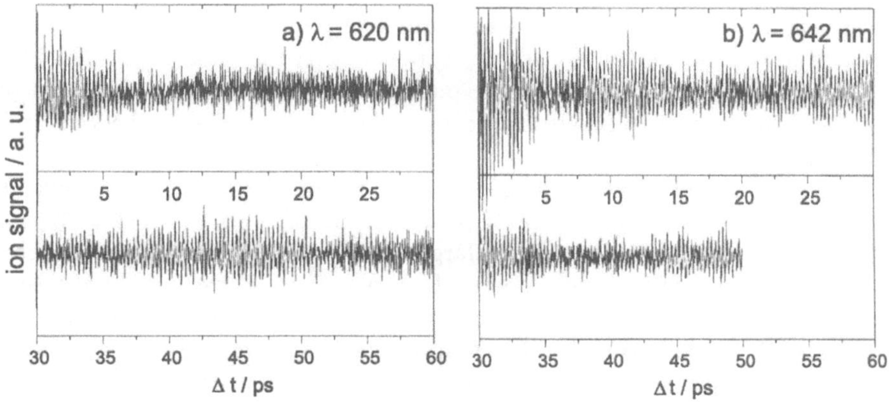

Fig. 3.21. Real-time spectra of 3PI signals of Na_2 excited with laser pulses of 110 fs duration with central wavelength (**a**) $\lambda = 620$ nm and (**b**) $\lambda = 642$ nm (taken from [333])

In both real-time spectra a fine oscillatory structure with an oscillation period $T_{A,620} \approx 312$ fs for 620 nm and $T_{A,642} \approx 297$ fs 642 nm, dominates the real-time evolution of the ion signal. For the shorter wavelength the first maximum of the ion signal appears after a delay time between the pump and probe pulses of$\Delta t_{\Phi,620} \approx 315$ fs, which is nearly equal to the measured oscillation with $T_{A,620} \approx 312$ fs, and the respective time for 642 nm is $\Delta t_{\Phi,620} \approx 160$ fs, approximately half of the corresponding oscillation period $T_{A,642}$.

As for K_2, the oscillation in intensity of the ion signal represents the propagation of a wave packet on the electronic $A\,^1\Sigma_u^+$ state. At $\Delta t = 0$ this wave packet is prepared by the pump pulse at the inner turning point of the A state well. Similarly to the case of the potassium dimer excited to its A state, an intermediate Rydberg state acts as a ladder in the ionization process at a certain internuclear distance. For the excitation wavelength 620 nm this Franck–Condon window is realized by the $(2)\,^1\Pi_g$ state. Gerber and coworkers [28] found a similar result in experiments with higher laser pulse power. Performing a difference potential analysis (see process (a) in Fig. 3.20) [334] easily proves this fact.[2] Oppositely to the analogous case in K_2, here the

[2] See also the excellent theoretical work of Machholm and Suzor-Weiner [335] on the control of Na_2^+ photodissociation by intense femtosecond laser pulses.

ionization is favored at internuclear distances lying close to the inner turning point of the wave packet propagation. This is nicely indicated by the delay time $\Delta t_{\Phi,620}$, amounting to one vibrational period.

In contrast, for 642 nm the corresponding delay time $\Delta t_{\Phi,642}$ amounts approximately to half of a vibrational period. The ionization with the longer wavelength is favored at the outer turning point of the wave packet's propagation. The intermediate gerade state responsible for the two-photon ionization step can also be identified as the $(2)\,^1\Pi_g$ state by a difference potential analysis. In the internuclear distance range of this transition to the ion the potential well of the $(2)\,^1\Pi_g$ state is rather flat (see process (b) in Fig. 3.20) [336]. Therefore, this state will be populated weakly during the ionization of the A state, and consequently no wave packet dynamics on the Π state's PES are expected.

A pronounced wavelength dependence of the two real-time spectra is present in the 'long-time' evolution of the pump&probe data. For 620 nm the oscillation amplitude, which is large close to the zero of time, disappears for intermediate delay times between 15 ps and 35 ps. Subsequently, it passes through a strong recurrence around $\Delta t_{r,620} \approx 47$ ps (see Fig. 3.21 a). Since the dimer's PES is anharmonic, the components in the wave packet get out of phase after a propagation time of about 15 ps and re-interfere around 47 ps. A similar recurrence time for a wavelength of 620 nm has been found before [26] and could be well simulated theoretically using a quantum dynamical ansatz [34–36].

A totally different behavior is observed, when pumping the A state with 642 nm radiation (Fig. 3.21 b). The fast oscillation described above is superimposed on a beat structure with a period of $T_s \approx 14$ ps. A revival already appears at $\Delta t^{(1)}_{r,642} \approx 11$ to 16 ps. Furthermore, recurrence phenomena appear between 25 ps and 30 ps and near 50 ps. To explain the deviations between the two pump&probe experiments we calculated the frequency components of the real-time data by a Fourier analysis. The resulting frequency components were compared with the energy spacings $\Delta w_{k,l}$ between two vibronic eigenvalues k and l found by energy-resolved spectroscopic methods. A frequency component $\omega_{k,l}$ corresponds to the vibrational period $T_{k,l} = 1/c\omega_{k,l}$ in appropriate units, where c is the speed of light.

The frequency components involved in the real-time spectra were extracted by a Fourier analysis of the normalized pump&probe data (see Fig. 3.22 a for excitation with $\lambda = 620$ nm, and Fig. 3.22 b for 642 nm). Both Fourier spectra are dominated by a band of frequencies centered around $107.8\,\mathrm{cm}^{-1}$ and $111.3\,\mathrm{cm}^{-1}$ for the respective wavelengths. This main frequency band contains the frequencies of the energy spacings of neighboring vibronic levels (i.e. $\Delta v = 1$), simultaneously excited by the spectrally broad femtosecond laser pulse. An additional frequency group with lower amplitude appears for both experiments, around $214\,\mathrm{cm}^{-1}$ in Fig. 3.22 a and around $222\,\mathrm{cm}^{-1}$ in Fig. 3.22 b. This contribution appears to be due to the fact that

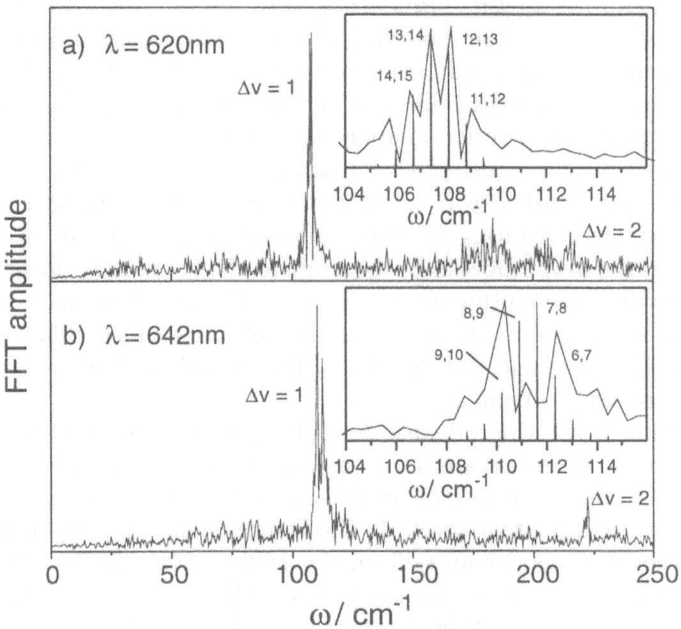

Fig. 3.22. Fourier spectra of the real-time 3PI signals of Na_2 excited with a 110 fs pump pulse at **(a)** $\lambda = 620$ nm and **(b)** $\lambda = 642$ nm (taken from [333]). The insets present the respective dominant frequency band in detail. The lines indicate the position as well as the approximate oscillator of the frequencies as expected from RKR calculations. The pairs of numbers identify the lines as level spacings $\Delta v = k - l$ of neighbored excited vibrational states with quantum numbers k and l

the laser pulse is also capable of coherently exciting vibronic levels with a difference in the quantum number of $\Delta v = 2$.

Further frequencies can be observed in Fig. 3.22 a. A rather weak frequency is present at $\omega \approx 90\,\mathrm{cm}^{-1}$, and stronger components are located around $\omega \approx 180\,\mathrm{cm}^{-1}$. These patterns are assigned to a wave packet propagating on the $(2)\,^1\Pi_\mathrm{g}$ PES around $v = 17$. This vibronic level can be directly excited from the $v = 0$ level of the ground state PES by two photons of the pump laser ([336]). In this case, the component at $90\,\mathrm{cm}^{-1}$ equals the energy spacing of neighboring vibronic levels of the Π state. A wave packet initiated by two photons of the pump pulse in the $(2)\,^1\Pi_\mathrm{g}$ state will be excited by the probe pulse to a doubly excited state at the outer turning point of the molecule's vibration [331]. By autoionization of this doubly excited state this process can – similarly to the findings of Gerber's group – produce a clearly recognizable effect in the ion signal. The doubled frequency ($180\,\mathrm{cm}^{-1}$) can be directly seen in the ion signal of Fig. 3.21 a at a delay time of about 20 ps, where a 184 fs vibration dominates the signal. This doubled frequency is an expression of a fractional revival in the $(2)\,^1\Pi_\mathrm{g}$ state (In the classical limit, here, the two components of the wave packet have a phase difference of π).

This enables an excitation to the autoionizing state at the outer turning point at twice the frequency of the molecule's classical vibration [54]. In the experiment at 642 nm no frequencies of similar origin have been detected.

The frequency group for $\Delta v = 1$ is illustrated in the insets of Fig. 3.22. The inset of Fig. 3.22 a shows the Fourier components in the real-time spectrum for 620 nm compared with the energy-level spacings $\Delta v = 1$ excited by the femtosecond laser. Each frequency is calculated from the RKR energies of [332]. The amplitude of a resulting frequency component is assumed to be proportional to the laser amplitudes at both energy levels. For each frequency the corresponding vibronic quantum number is given. The distribution of the frequencies involved is very smooth. It reflects the excitation of the vibronic levels of the slightly anharmonic A state PES by the femtosecond laser pulse with its nearly Gaussian spectrum.

In the inset of Fig. 3.22 b the corresponding plot is presented for 642 nm. Here, the RKR-based frequencies are not in agreement with the frequencies obtained from the 3PI real-time spectra. In the center of the frequency group the deviations are especially pronounced. In the Fourier spectrum, two main frequencies, both located slightly away from the center of the $\Delta v = 1$ frequency group, dominate the pattern. These frequencies lie at $\omega_1 = 112.4\,\mathrm{cm}^{-1}$ and $\omega_2 = 110.2\,\mathrm{cm}^{-1}$. The classical beat period from the two frequencies

$$\omega_\mathrm{s} = \omega_1 - \omega_2 = 2.2\,\mathrm{cm}^{-1}, \tag{3.24}$$

$$T_\mathrm{s} = \frac{1}{\omega_\mathrm{s} c} = 15.15\,\mathrm{ps} \tag{3.25}$$

gives the beat oscillation time as is observed in the real-time spectrum in Fig. 3.21 b.

To explain the position of the frequencies in the Fourier spectrum one has – similarly to the case of K_2 – to consider that nearly every vibronic level of the A state is perturbed by a vibronic eigenstate of the $\mathrm{b}\,^3\Pi_\mathrm{u}$ state [320]. This perturbation induced by SOC results in a shift of the respective level of the A state and, therefore, the energy-level spacings to neighboring levels are affected.

A positive energy shift of a level will enlarge the spacing from the lower-lying levels and the difference from higher levels will become smaller. The perturbation of each level in the A state is strong for values of the rotational quantum number $J \gg 20$, except for $v = 8$ where the J values for a strong perturbation lie between 8 and 14 [320]. For this vibronic level the shift in energy due to the SOC process was not given by Effantin et al. [320].

Next, the energy-level shifts found in the experiment will be analyzed. Following the ideas of [320], appropriate energy-level shifts to the RKR values of [332] have to be introduced. In the molecular beam the rotational temperatures for the molecules are smaller than 15 K. Hence, the maximum of the rotational population in the ground state is given by $10 < J_\mathrm{max} < 20$. The selection rules for the transition to the A state indicate that this maximum remains more or less unchanged. Therefore, a SOC-induced shift of the

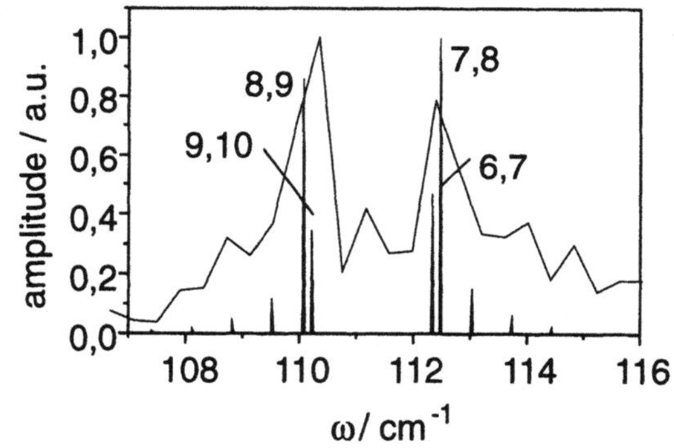

Fig. 3.23. Part of Fourier spectrum of the real-time 3PI signal of Na_2 obtained for excitation with λ = 642 nm (taken from [333]). The lines indicate the position as well as the approximate oscillator strength of the frequencies expected from RKR calculations. The pairs of numbers identify the lines as level spacings $\Delta v = k - l$ of neighboring excited vibrational states with quantum numbers k and l

$v = 8$ level in the A state is expected. An energy-level shift for this vibronic RKR eigenstate is fitted to the position of the energy-level spacings calculated by the Fourier analysis of Fig. 3.22. In Fig. 3.23 the Fourier spectrum is compared with the RKR frequencies given by [332]. An energy level shift for $v = 8$ of $\Delta E_8 = 0.85\,\mathrm{cm}^{-1}$ can be estimated. The result of this procedure, summarized in Table 3.2, shows very good agreement of the resulting energy spacings with the Fourier components.

Table 3.2. Vibrational RKR level pairs reached by the excitation laser and corresponding energy spacings without and with energy-level shift for $v = 8$ of $0.85\,\mathrm{cm}^{-1}$, compared with experimental and theoretical found for Na_2

k, l	$\omega_{RKR}/\mathrm{cm}^{-1}$ [a]	$\omega_{shift}/\mathrm{cm}^{-1}$ [b]	$\omega_{FFT}/\mathrm{cm}^{-1}$ [c]
5, 6	113.031	[d]	[e]
6, 7	112.327	[d]	112.4 (6,7 and 7,8)
7, 8	111.622	112.472	
8, 9	110.971	110.067	110.1 (8,9 and 9,10)
9,10	110.215	[d]	
10,11	109.512	[d]	[e]
11,12	108.810	[d]	108.9
12,13	108.108	[d]	108.1
13,14	107.408	[d]	107.4
14,15	106.708	[d]	106.7
15,16	106.009	[d]	[e]

[a] Vibrational level spacings from [332].
[b] Level spacings with introduced shifts $\Delta E = 0.85\,\mathrm{cm}^{-1}$ for $v = 8$.
[c] Fourier components calculated from real-time data.
[d] Not changed.
[e] Not observed.

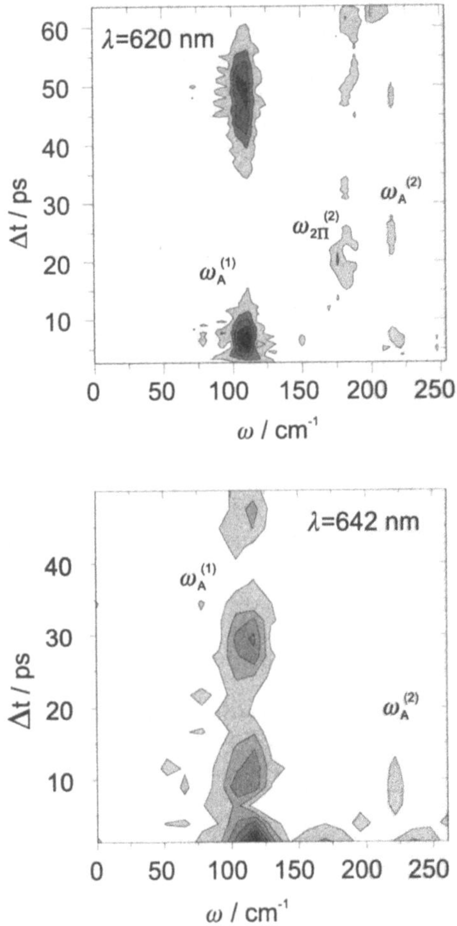

Fig. 3.24. Logarithmic contour plots of the spectrograms $I(\Delta, \omega)$ of Na_2 excited to its A state for two different excitation wavelengths λ (taken from [262]). The Fourier amplitude increases from white to black

At the end of this section spectrograms of both real-time spectra are presented. They neatly summarize what has been discussed above and make clear the interplay of the dimer's various modes. The logarithmic contour plots of the spectrograms are presented in Fig. 3.24. The Fourier amplitude is indicated by increasing gray from white to black. The frequency bands already seen in the frequency spectra (Fig. 3.22) can now be examined with their temporal evolution.

This experiment on Na_2 again nicely indicates that femtosecond spectroscopy is a highly sensitive method to investigate the perturbation of induced wave packet propagation. Energy shifts of vibrational levels of excited electronic states can be detected with a resolution of better than $0.1\,\text{cm}^{-1}$. This enables the estimation of the strength of spin–orbit coupling between crossing electronic states. The shift causes the observable changes in the revival structures of the two real-time spectra. Besides this, the varying wave packet dynamics observed for perturbed and unperturbed cases have been analyzed in great detail, revealing different transition pathways for excitation at 620 nm and 642 nm. The transition pathways of the (1+2)-photon process can be directly deduced from the phase of the real-time spectra. For $\lambda_{\text{pump}} = 620\,\text{nm}$ the two-photon step takes place at the inner turning point of the wave packet in the A state; for $\lambda_{\text{pump}} = 642\,\text{nm}$ it happens at the outer turning point. The wave packet dynamics of the $(2)\,^1\Pi_g$ state, acting as a Franck–Condon window in the two-photon ionization step, is seen while exciting the dimer at 620 nm.

3.1.5 Multiphoton Ionization Processes in K_2: from Pump&Probe to Control

As shown before, the interplay of theory and experiment enables an excellent understanding of laser-pulse-induced MPI dynamics of small molecules. Beyond this, it is now of great interest to develop new, far-reaching concepts that use these special dynamics of coherent superpositions of states prepared by femtosecond laser pulses. The main goal is to influence a molecular system, either to focus on vibrational modes in selected potential-energy surfaces or to guide the system into distinctive reaction channels. The control of reaction channels by femtosecond pulses was first suggested theoretically by Tannor, Kosloff, and Rice [61, 337] and was later also observed in experiments by Zewail, Gerber, et al. [66, 338]. In both cases the delay time between the pump and probe (control) laser pulses was the control parameter. With some modifications of the powerful pump&probe technique it is possible to drive a molecular wave packet to a desired location on the PES, from which selective dynamic processes (e.g. chemical reactions [66, 338]) can be initiated.

An excellent candidate to investigate the principles of a control experiment would be a simple molecule, which is relatively easy to handle, both theoretically and experimentally. Here, the potassium dimer is chosen as a suitable molecular system. Consistent experimental and theoretical investigations demonstrate that the intensity of the initial laser field will be the tool for control of the observed dynamics. As an extension to the strategy of Tannor, Kosloff, and Rice it is shown here that different laser intensities induce different transition pathways (Fig. 3.25), e.g. the competing pathways of direct MPI via electronic excited states and indirect MPI via resonant impulsive stimulated Raman scattering (RISRS) into the molecule's ground state [67, 339–346]. In contrast to emission spectroscopy, where a continuous wave laser or a nanosecond laser pulse is used to populate specific vibrational eigenstates of the initial (here electronic ground) state [347], the RISRS processes prepare a coherent vibrational state within the pump pulse duration which evolves on the ground state surface and whose vibrational motions can now be detected in the transient ion signal. The control mechanism reported here is in line with other work on similar subjects [33, 67, 294, 348–350], and the investigation on K_2, in particular, leads to new aspects compared to Na_2. Complementary aspects have already been reported for Cs_2 [37].

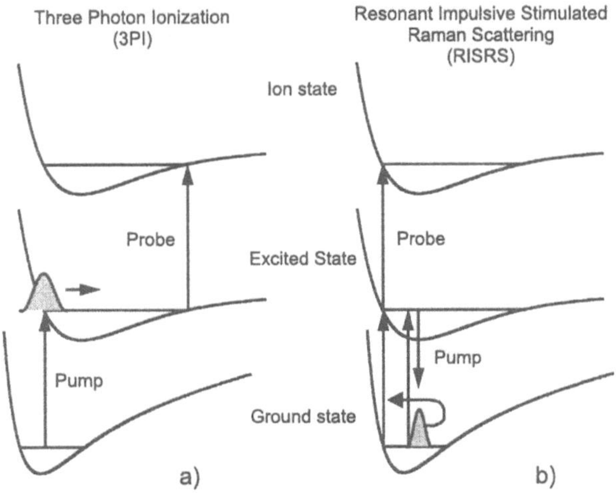

Fig. 3.25. Schematic diagram for **(a)** 'direct' two-photon ionization and **(b)** 'indirect' multiphoton ionization via RISRS processes (taken from [328])

The experimental and theoretical femtosecond real-time studies presented here were performed for moderate and high laser intensities. The theoretical approach to these investigations is given in Sect. 2.2.1 and further details can be found in [42, 295, 328, 351]. The experiments were performed at a wavelength of 840 nm. Hence, again a (1+2) pump&probe cycle was used. The width of the pump pulse was in both cases 60 fs. For the moderate excitation a peak power of the pump pulse was estimated to be 0.5 $GWcm^{-2}$ increased by a factor 10 for the high-power case.

A comparison of experimental and theoretical total ion signals for delay times up to 10 ps is presented in Figs. 3.26 and 3.27 for moderate and high laser intensities respectively. The first obvious difference is that the transient ion signal for moderate laser intensity shows a distinct oscillation, whereas for high-laser-field excitation the oscillatory structure is rather poorly resolved. The disturbed oscillatory structure might be due to interferences of different wave packet signals.

To elucidate the ionization mechanism for the two selected laser intensities the relevant snapshots of the transition processes are presented in Figs. 3.28 and 3.29. The actual wave packets are represented by the absolute values of their amplitudes. In both cases the delay time between the pump and probe pulses was 300 fs, the time period in which the wave packet of the $A\,^1\Sigma_u^+$ state is located close to its outer classical turning point. In each case only the wave packets which participate in the dominant pathway are shown. In the present four-state-plus-ion-continuum model two main transition pathways can be distinguished. Ionization pathway (a) involves the electronic states $X\,^1\Sigma_g^+$, $A\,^1\Sigma_u^+$, $2\,^1\Pi_g$, and the ion ground state continuum whereas ionization

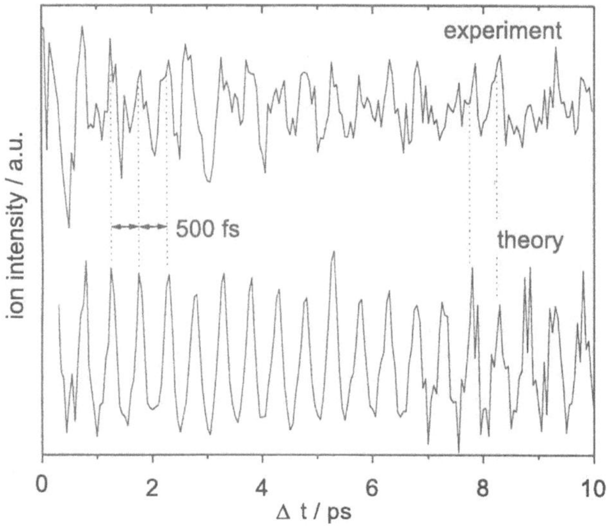

Fig. 3.26. Comparison between the measured and calculated real-time dynamics in $^{39,39}K_2$ at moderate laser pulse intensities, adapted from [42, 351]. The fluctuations in the experimental pump&probe signal are mainly caused by instabilities of the molecular beam during the experiment. The ratio of the modulation amplitude to the overall averaged ion intensity amounts to 0.6 (expriment) and 1.9 (theory). The difference is due to the fact that, in contrast to the theoretical data, the experimental ion signal contains a constant fraction of ions which originate from the interaction of a single pump or probe pulse with the molecules (taken from [328])

pathway (b) involves the electronic states $X\,^1\Sigma_g^+$, $A\,^1\Sigma_u^+$, $4\,^1\Sigma_g^+$, and the ion ground state continuum. In Fig. 3.30 the corresponding norms of the states involved, as well as the total norm, are presented for the same two selected laser intensities.

In the case of moderate laser fields detailed calculations have shown that the only efficient ionization pathway is pathway (a). According to [42, 351], the pump pulse prepares a wave packet in the $A\,^1\Sigma_u^+$ state (Fig. 3.28 a), centered energetically at the vibrational level $v = 11$. The $2\,^1\Pi_g$ state is populated to only a small extent with a wave packet centered at $v = 3$. The potential-energy gap Δ_{A2} of 1.494 eV at the outer ($r = 5.4$ Å) and of 1.576 eV at the inner ($r = 4.35$ Å) turning points of the $A\,^1\Sigma_u^+$ state wave packet are of decisive importance in this process. Even taking into account the energy width due to the rather short pump pulse, it is exclusively the potential-energy difference at the outer turning point which is in good resonance with the laser frequency of 840 nm (≙1.475 eV). Therefore, significant transitions into the ion ground state are possible only while the $A\,^1\Sigma_u^+$ state wave packet is at its outer turning point. Now the probe pulse can ionize the molecule using the $2\,^1\Pi_g$ state as a resonant state. This process is indicated in Figs .3.28 b,

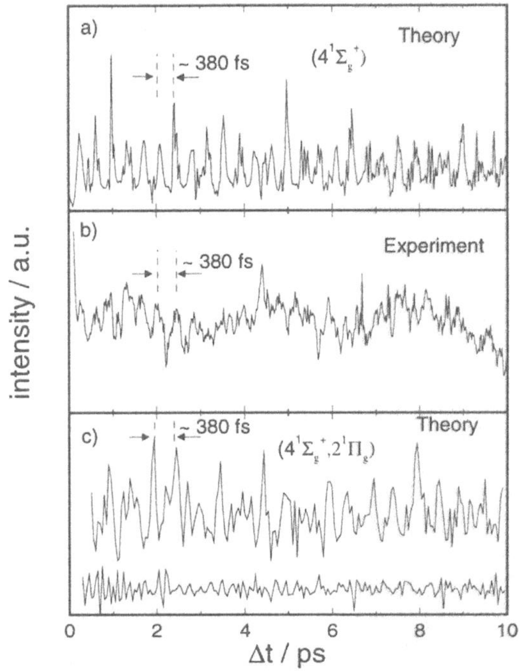

Fig. 3.27. Comparison between the measured and calculated dynamics in $^{39,39}K_2$ at high laser pulse intensities. **(a)** The ion population when only the $4\,^1\Sigma_g^+$ state is included in the calculation, **(b)** the experimental signal, **(c)** the resulting ion population (*upper curve*) when both states $4\,^1\Sigma_g^+$ and $2\,^1\Pi_g$ are included in the calculation, together with the interference term $2\mathrm{Re}\langle\Psi(E_k)|\Psi(E_{k'})\rangle$ (*lower curve*) of both ionization pathways in the ion continuum. The fluctuations in the experimental pump&probe signal are mainly caused by instabilities of the molecular beam during the experiment. The ratio of the modulation amplitude to the overall averaged ion intensity amounts to 0.6 (experiment) and 1.1 (theory). The difference is due to the fact that, in contrast to the theoretical data, the experimental ion signal contains a constant fraction of ions which originate from the interaction of a single pump or probe pulse with the molecules (taken from [328])

c which show the situation around the maximum of the probe pulse. The transition occurs, as indicated by the arrows, at the outer turning point (pathway (a) above) of the wave packet in the $A\,^1\Sigma_u^+$ state. In summary, a 'direct' pure (1+2) multiphoton process into the ion ground state takes place in the case of moderate laser intensities. In this context 'pure' means that the first step is a pure 1 photon process, i.e. the $2\,^1\Pi_g$ state does not need to be populated, and the second step, induced by the probe laser, is a pure two-photon excitation from the $A\,^1\Sigma_u^+$ state via the $2\,^1\Pi_g$ state into the ion continuum. The snapshots correspond to a calculation for one optimally selected kinetic energy $E_k = 0.263\,\mathrm{eV}$, which is indicated by the shorter arrow in Fig. 3.28 b.

In the transition process for higher laser fields, the contribution of the $4\,^1\Sigma_g^+$ state becomes important. To illustrate this process let us first consider the simulation when the $4\,^1\Sigma_g^+$ state, but not the $2\,^1\Pi_g$ state, is included in the calculation (see Fig. 3.29). Here, the pump pulse prepares wave packets both in the $A\,^1\Sigma_u^+$ and in the $4\,^1\Sigma_g^+$ states (Fig. 3.29 a), i.e. a two-photon excitation takes place. Simultaneously, a significant motion of the wave packet on the electronic ground state surface (centered at $v = 1$) is induced by efficient RISRS processes from the excited states (mainly from the $A\,^1\Sigma_u^+$ state), as indicated schematically in Figs. 3.25 b and 3.29 a. Again the wave packet of the $A\,^1\Sigma_u^+$ state is at its outer turning point when the probe pulse is turned on (Fig. 3.29 b). Simultaneously, the wave packet in the $4\,^1\Sigma_g^+$ state has moved even further away from its inner turning point. Figures 3.29 b, c show the situation of the dynamics around the maximum of the probe pulse. Owing to the enhanced (compared to moderate laser fields) amplitude of motion on the ground state surface, a transition pathway at the inner turning point is now opened (pathway (b)). The turning point of the ground state wave packet is shifted to shorter bond lengths, from 3.82 Å to 3.75 Å which now corresponds well to the inner turning point of the wave packet in the $4\,^1\Sigma_g^+$ state (3.76 Å). This opens a new pathway (b), different from pathway (a), and an efficient direct resonant three-photon transition (Fig. 3.29 b) at the inner turning point of the wave packet in the $X\,^1\Sigma_g^+$ ground state takes place. The pathway (b) can be enhanced by two- and one-photon transitions from the $A\,^1\Sigma_u^+$ and $4\,^1\Sigma_g^+$ states respectively. A third conceivable enhancement via the inner turning point of the $2\,^1\Pi_g$ state wave packet at 4.35 Å remains inactive, because here the energy gap Δ_{A2} stays out of optimal resonance, as in the case of moderate laser intensities. The snapshots are taken for one optimally selected kinetic energy $E_k = 0.209$ eV of the ionic continuum, which is indicated by the short arrow in Fig. 3.29 b.

The increasing role of the excitation scheme of pathway (b) at higher laser intensities reflects the intensity-dependent enhancement of the RISRS process. Basically, higher laser intensities just stimulate more Raman processes and their effect on the ground state wave packet is reinforced. Similar investigations have been reported by Banin et al. [67] for a two-electronic-surface system. These authors pointed out the significance of $N\pi$ ($N = 1, 2...$) pulse conditions for the RISRS process. The π pulse is defined [352] by

$$\frac{1}{2} W t_f = \pi, \tag{3.26}$$

where t_f is the pulse duration and

$$W \gg \mu(r) \frac{1}{\hbar t_f} \int_0^\infty E(t')\mathrm{d}t'. \tag{3.27}$$

In Fig. 3.12 for various electronic states of K_2 the electronic transition dipole moment μ is shown as a function of the bond length r.

According to [67], the π-pulse condition creates a dynamical 'hole' in the position distribution of the ground state at the point of resonance, while

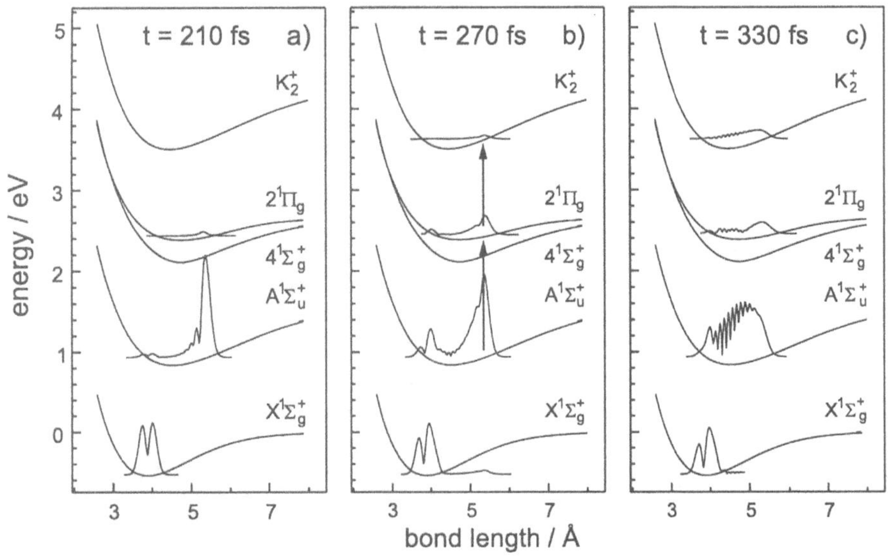

Fig. 3.28. Snapshots of the wave packets in their respective electronic states for a delay time of 300 fs between the pump and probe lasers for the (direct) (1+2) multiphoton ionization process in K_2 (moderate laser field). The wave packets are represented by their absolute values. The two-photon step is indicated by the arrows. The $2\,^1\Pi_g$ state is included in the dynamical calculation; the $4\,^1\Sigma_g^+$ state is inactive (taken from [328])

an increase in laser intensity leads to faster Rabi cycling and the 2π-pulse cycles the amplitude back to the ground state, inducing there a momentum kick to the 'hole'. Thus for a pulse with the proper integrated fluence of 2π the contribution from the excited states' dynamics to the transient ion signal can be minimized and the motion of the coherent vibrational state in the ground state can be maximized. These considerations are adapted here to explain the intensity-dependent enhancement of the RISRS process in the present high- and moderate-power pump&probe experiments for the potassium dimer. Here, up to four electronic states plus the ion continuum can be involved. However, to discuss the effect of the RISRS process on the ground state dynamics we can focus our considerations exclusively on the dominant $X\,^1\Sigma_g^+$–$A\,^1\Sigma_u^+$ interaction. For this subsystem the present high-power pulse yields

$$\frac{1}{2}Wt_f \gg 4.14 \gg 1.3\pi. \tag{3.28}$$

For this estimate, the transition dipole moment

$$\mu_{XA}(r) = \mu_{XA}(r_e) = 4.5ea_0 \tag{3.29}$$

was used. Thus the higher laser pulses employed exceed the π-pulse condition. Accordingly, the ground state population after the pump pulse is nearly

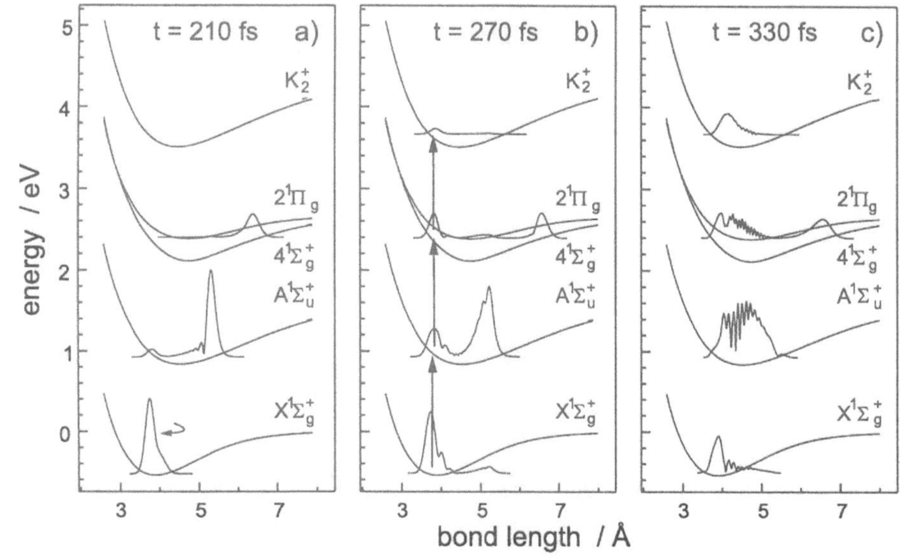

Fig. 3.29. Snapshots of the wave packets in their respective electronic states for a delay time of 300 fs between the pump and probe lasers are shown for the (indirect) (2+3) multiphoton ionization process in K_2 (high laser field). The wave packets are represented by their absolute value. The effect of the RISRS process is indicated by the curved arrow. The subsequent three-photon step is indicated by the arrows. The $4\,^1\Sigma_g^+$ state is included in the dynamical calculation; the $2\,^1\Pi_g$ state is inactive (taken from [328])

50% (see Fig. 3.30 b) and the Rabi cycling during the pump pulse can be recognized. A large-amplitude motion on the ground state surface is induced (see Fig. 3.29 a). In contrast for moderate intensities the present laser pulse is estimated to be 0.62π, i.e. a fraction of a π pulse. Thus a 'hole' is created in the ground state at the domain of resonance (see Fig. 3.28 a), but without an efficient momentum kick, and only 31% (see Fig. 3.30 a) of the total population stays in the ground state. Thus the dynamics of the excited A state are dominant.

The ionization pathways (a) and (b) are confirmed by comparison of the temporal evolutions of the theoretical and experimental ion signals (see Figs. 3.26 and 3.27). In the case of moderate laser intensities (Fig. 3.26) the experimental and theoretical curves show strong oscillations with a period of 500 fs. They correspond to the oscillation periods of the induced wave packet in the $A\,^1\Sigma_u^+$ state [42, 351]. Measurement and theory are in good agreement in terms of the position of the maxima and the envelope intensity modulation. Sharp variations in the ion signal intensity occur in the simulation for delay times larger than 6 ps. In the same time range the noise in the experimental data increase. Since no noise exists in the simulation, the sharp beat structure might originate from wave packet interferences, as reported

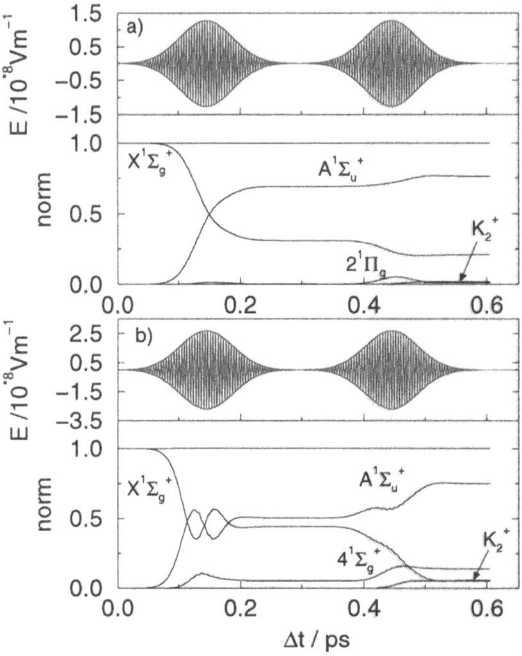

Fig. 3.30. Changes of population of the electronic states of K_2 involved in the MPI processes, together with the pump and probe laser pulses for a delay time of 300 fs (taken from [328]). Panel **(a)** shows the situation for laser pulses of moderate intensities and panel **(b)** for high laser intensities. The creation of a ground state wave packet by RISRS processes, as well as the Rabi-type process, is seen for the more intense laser pulse in **(b)**

e.g. in [37]. In the case of moderate laser intensities for which the transition paths of the preparing pump and ionizing probe laser photons are located at different turning points (see Fig. 3.28), these interferences can only occur when the wave packet in the $A\,^1\Sigma_u^*$ state begins to spread. The existence of only one ionization pathway of type (a) in the case of moderate laser fields is reflected in the pure oscillations of the experimental and theoretical total ion signals. No interference with other pathways occurs. The findings are strongly supported by the Fourier transform of the experimental ion signal [42]. The Fourier spectra, presented in Sect. 3.1.3, exhibit a strong, dominant band in the region of approximately 66 cm^{-1}, belonging to the $A\,^1\Sigma_u^*$ state, and a very weak single line at about 90 cm^{-1} attributed to the ground state.

High laser intensities change the dynamical situation. Now, in principle two ionization pathways, i.e. (a) and (b), are accessible. The oscillation period of 380 fs in the experimental signal (Fig. 3.27b) and in the theoretical simulation (Figs. 3.27 a, c) corresponds to the vibrational motion of the wave packet in the ground state induced by the RISRS process and indicates that ionization pathway (b) is now dominant. Again a structure of sharp modulations is superimposed on the ion signal. The interference between wave packets prepared by the pump and probe laser pulses is now more pronounced because the ionization pathway (b) is located at the inner turning point for both pulses. In Fig. 3.27 a the ionization pathway (b) is considered exclusively in the calculation. This simulation already reproduces correctly the positions of the maxima in the pump&probe spectrum. The Fourier transform of the experimental ion signal [46] shows its most prominent peak at the frequency of the ground state and therefore also supports the interpretation that the dynamics of the ground state are dominantly reflected in the transient ion signal. The spectroscopic properties of K_2, i.e. the possibility of the two different ionization pathways for the vibrational motion, in the $A\,^1\Sigma_u^+$ state via (a) and in the $X\,^1\Sigma_g^+$ state via (b), localized at the outer and inner turning points, respectively, make it possible to detect the RISRS process in the transient ion signal.

Another aspect of the high-power pump&probe spectrum is the experimentally observed beat structure of 2.6 ps (Fig. 3.27 b). Again the existence of two different ionization pathways is of decisive importance. The vibrational periods of the excited potential surfaces involved ($A\,^1\Sigma_u^+$, $4\,^1\Sigma_g^+$, and $2\,^1\Pi_g$) lie in the range 500 – 620 fs. Their beat frequency with the ground state oscillations of 380 fs has a period of 1.3 ps. Thus the experimentally observed beat frequency of 2.6 ps can only be caused by the influence of two additional ionization pathways. These may be located, for instance, at the inner and outer turning points, mirroring half the oscillation period of the wave packet in one of the excited states. The two states $4\,^1\Sigma_g^+$ and $2\,^1\Pi_g$ in combination just open the possibility of mirroring half the oscillation period of the wave packet in the $A\,^1\Sigma_u^+$ state, i.e. the first-harmonic (via the $4\,^1\Sigma_g^+$ state at the inner turning point) and second-harmonic (via the $2\,^1\Pi_g$ state at the outer turning point) transitions from the $A\,^1\Sigma_u^+$ state into the ion continuum. Therefore, the experimentally observed beat structure with a period of about 2.6 ps in the temporal evolution of the ion signal cannot be reproduced in the calculated signal when only the $4\,^1\Sigma_g^+$ state is included in the matrix representation of the time-dependent Schrödinger equation (see Fig. 3.27 a).

More extensive calculations have been performed taking into account both pathways simultaneously. The resulting coherent sum of the continuum contributions is shown in the upper curve of Fig. 3.27 c, together with the interference term

$$2\mathrm{Re}\langle\Psi_{i,k}(E_k)|\Psi_{i,k'}(E_{k'})\rangle \tag{3.30}$$

(lower curve of Fig. 3.27 c) of the two dominant contributions ($E_k = 0.263\,\mathrm{eV}$ for pathway (a) and $E_{k'} = 0.209\,\mathrm{eV}$ for pathway (b)). Comparison with the experimental curve clearly demonstrates that the intensity modulation observed in the experiment is due to the interference terms of the continuum. The neglect of the interference term in the calculation results in an ion signal which reflects dominantly the vibration in the $A\,^1\Sigma_u^+$ state, i.e. the 500 fs oscillation period. Besides this, it was found that the total ion signal is sensitive to the relative strength of the two ionization pathways, with the conclusion that they should be of similar strength. This finding is also supported by the Fourier spectrum of the experimental data [46], which shows next to the highest peak, at the frequency of the $\mathrm{X}\,^1\Sigma_g^+$ state, a strong peak at the frequency of the $\mathrm{A}\,^1\Sigma_u^+$ state. Therefore, not only the transition dipole moments between the neutral states but also the transitions $4\,^1\Sigma_g^+ \rightarrow$ ion continuum and, alternatively, $2\,^1\Pi_g \rightarrow$ ion continuum are important when several ionization pathways located at different positions contribute to the ion signal. The parameters assumed for the corresponding transition dipole moments may be the reason for the fact that the beat structure in the theoretical pump&probe spectrum is not as pronounced as in the experimental one. However, from the theoretical results we can conclude that in the case of high laser power the contributions from the ionization pathway (b) cause the oscillation period, the contributions from pathway (a) cause the broadening of the oscillations in the total ion signal, and the interference of both pathways, including the contributions from first- and second-harmonic transitions, are responsible for the beat structure.

The characteristics of the pump and probe pulses, namely wavelength, intensity, and pulse width, induce special dynamics by 'femtosecond state preparation' in the molecules. They define the potential curves involved, the energetic location of the wave packets on the PES, and consequently also the location of the turning points with respect to the nuclear coordinate. The positions of the turning points define the possible regions for optimal resonant transitions. For a given wavelength the variation of the laser field intensity influences the probability of multiphoton processes, the number of Rabi cycles, and the efficiency of RISRS processes [348]. Here, these effects are realized. During the transition from moderate to high laser fields the contributions from the energetically close-lying $4\,^1\Sigma_g^+$ and $2\,^1\Pi_g$ states influence decisively the resulting ionization pathways and consequently the total ion signal. In the case of moderate laser fields, only the pathway (a) via the $2\,^1\Pi_g$ state is accessible. More intense laser fields, however, induce a large-amplitude motion on the ground state surface by an efficient RISRS process. The oscillatory shift of the wave packet around its equilibrium position is now large enough that its inner turning point is shifted towards shorter bond lengths and reaches a nuclear conformation which now fulfills the resonance conditions with the laser frequency at the inner turning point via the $\mathrm{A}\,^1\Sigma_u^+$ and $^1\Sigma_g^+$ states, and opens the pathway (b). The $4\,^1\Sigma_g^+$ state thus becomes a resonant state in the three-photon excitation scheme. Its Franck–Condon (FC) window at the inner turning point in the $\mathrm{X}\,^1\Sigma_g^+$ state, however, is not selective. It is also open for transitions from the $\mathrm{A}\,^1\Sigma_u^+$ and $4\,^1\Sigma_g^+$ state, owing to the relative position of the potential curves. The dominant contribution to the total ion signal originates from the $\mathrm{X}\,^1\Sigma_g^+$ state's vibrational motion, but the frequencies of all the other states involved can be resolved in the Fourier spectrum [46].

The role of the $2\,^1\Pi_g$ state changes also during the transition from moderate to high intensities. In all cases, the $2\,^1\Pi_g$ state opens an important transition path. Its FC window is located at the outer turning point of the wave packet in the $A\,^1\Sigma_u^+$ state and allows an optimal resonant transition, which is the first step of the significant TPI process. In the case of moderate laser intensities this TPI signal reflects the motion of the wave packet in the $A\,^1\Sigma_u$ state. For high laser intensities, the same FC window opens the way to the second-harmonic transition from the $A\,^1\Sigma_u^+$ state, which then is reflected in the beat structure as a coherent superposition of the first and second harmonic of the $A\,^1\Sigma_u^+$ state vibrational motion and the ground state vibrational motion. By shaping a perfect 2π pulse it should be possible to make transition pathway (b) even more pronounced.

The possibility to choose between different pathways via different electronic states in the excitation ladder of K_2 opens the way from pump&probe to control spectroscopy. The analysis shows that under specific spectroscopic conditions, for fixed photon energy and pulse duration, the intensity of the laser field (as in other cases [33, 67, 348]) serves as a control parameter, which may switch on two different excitation mechanisms, i.e. MPI or RISRS plus MPI. The special molecular properties required are that the forms of the potential surfaces involved and the relative positions of their minima have to differ. The induced femtosecond dynamics can then probe selected molecular modes. Contrary to the lighter homologue Na_2, in K_2 these modes are reflected in the oscillation periods of the transient ion signal and not only in the Fourier spectrum [33]. This is rendered possible by the existence of the two ionization pathways (a) and (b). If both pathways were located at the inner turning point, neither the vibration of the wave packet in the $A\,^1\Sigma_u^+$ state for moderate intensities nor the vibration of the wave packet in the $X\,^1\Sigma_g^+$ state could be resolved independently in the pump&probe spectrum. Thus the effect of the RISRS process can, contrary to the case of Na_2 [33], be detected directly in the transient ion signal of K_2. In this respect the K_2 molecule can be regarded as a model system whose properties should also be found in larger systems with more degrees of freedom and reaction channels. One might state, therefore, that this example nicely demonstrates that femtosecond chemistry is providing exciting new discoveries in the field of control of chemical and physical processes. In the optimal case, reaction pathways can be selected to investigate special modes and/or to achieve product selectivity.

3.1.6 Fractional Revivals of Vibrational Wave Packets in the NaK $A\,^1\Sigma^+$ State

Compared with their homonuclear analogues, heteronuclear alkali molecules are less well characterized. Up to now only NaK has been subject to a femtosecond study [53, 353, 354]. Since wave packet dynamics is sensitive to the influence of any perturbation of the PESs involved, NaK with its lack of a symmetry center might be an even better candidate to study perturbed wave packet propagation. In the heteronuclear molecule NaK, similarly to K_2 and Na_2 the $A\,^1\Sigma^+$ state interacts with the crossing triplet state ($b\,^3\Pi$) by spin–orbit coupling. Therefore, NaK should reveal a real-time spectrum which is rich in different revival structures. To understand this, a few remarks on total and fractional revivals will be made here first. The long-term real-time spectra (> 100 ps) of the two isotopes $^{23}Na^{39}K$ and $^{23}Na^{41}K$ are subsequently presented and analyzed in the time and frequency domains. As seen before, the revival structures can be studied best by means of spectrograms $I(\omega, \Delta t)$, allowing the direct and intuitive visualization of e.g. fractional revivals.

Revivals of Molecular Wave Packets. On a harmonic PES a wave packet completely regains its initial shape after the period, owing to the equidistant character of the spectra of the states it is formed by. In this sense the wave packet dynamics may be interpreted by means of quantum beats among the coherently excited states.

Owing to anharmonicities of the PES, a wave packet normally spreads. As a first consequence the amplitude of the oscillation decreases with increasing time after preparation, i.e. a destruction of the wave packet appears. However, this spreading is not at all an irreversible one. In their detailed description of the long-term evolution of wave packets Averbukh and Perel'man [54, 355] discussed the phenomena of wave packet revivals and fractional revivals. At a wave packet revival the wave function of the prepared system can be subject to a certain sequence of reconstructions which provide regular, well-localized structures in the probability density. The form of each condensation of the probability density is directly determined by the shape of the initial wave packet. The revival time T_{rev} is given by

$$T_{rev} = 2T_{vib}\left(\hbar\left|\frac{\partial\omega}{\partial E}\right|\right)^{-1}, \tag{3.31}$$

with T_{vib} the classical vibrational period in the anharmonic potential $E(\omega)$. At T_{rev} the phase shifts between different energy components inside the wave packet become exactly integer multiples of 2π. Then the evolution of the initial wave packet restarts. Since the revival structure depends strongly on $\partial\omega/\partial E$, it enables a sensitive characterization of the local anharmonicity of the excited $A\,^1\Sigma^+$–$b\,^3\Pi$ system. Besides this, for any irreducible rational fraction m/n the initial wave packet splits at a time close to

$$T_{frev}^{m,n} = \frac{m}{n}T_{rev} \tag{3.32}$$

into r ($r = n/2$ for even n and $r = n$ for odd n) spatially separated wave packet fractions performing a periodic evolution with relative time shift T_{vib}/r with respect to each other [54]. Averbukh and Perel'man could demonstrate this for atomic Rydberg states excited with laser pulses of 10 ps duration, measured by Parker and Stroud. [356]. In molecular systems fractional revivals have been observed in Na_2 and K_2 [262, 333] and in halogen dimers [51, 327].

Real-Time Dynamics of Excited NaK Molecules. Berg et al. measured the wave packet dynamics of NaK excited to its $A\,^1\Sigma^+$ state up to $\Delta t = 40$ ps [353]. They used a heat-pipe oven to produce a rather hot ensemble of NaK molecules in the electronic ground state. With an 85 ps laser pulse at 790 nm a wave packet was prepared on the A state PES. The wave packet's propagation was followed by a delayed probe pulse, which induced an excitation of the $E\,^1\Sigma^+$ state. By detecting the fluorescence of the E state, the real-time evolution of the wave packet could be recorded within the first 40 ps.

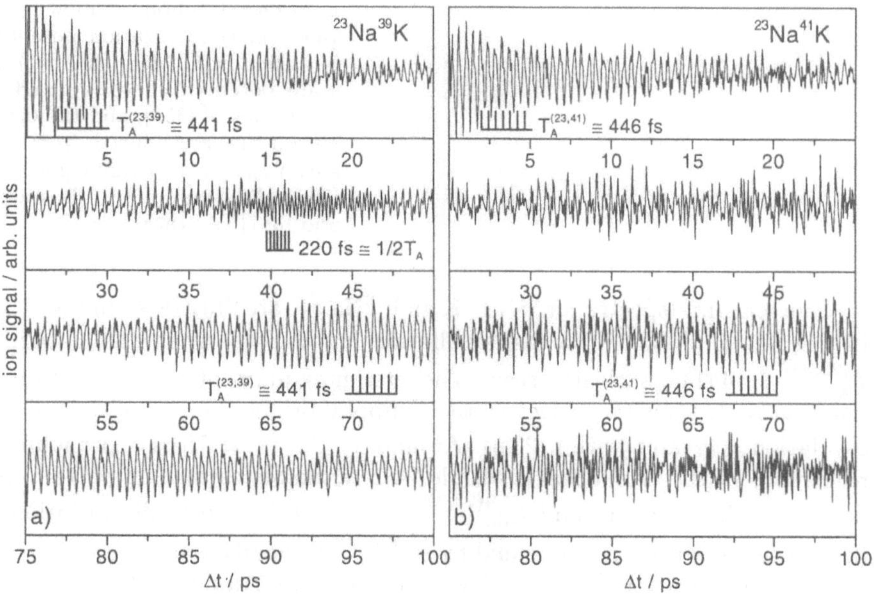

Fig. 3.31. Real-time spectra mirroring the wave packet dynamics of the $A\,^1\Sigma^+$ state for the isotopes $^{23}Na^{39}K$ (**a**) and $^{23}Na^{41}K$ (**b**). T_A is the vibrational period of the wave packet in the A state (taken from [53])

Here, long-time dynamics of an initially extremely cold ensemble of NaK molecules excited to its electronic A state are presented. This will allow a detailed observation of revival structures. As a rough estimate, the occurrence of the total revival can be calculated from spectroscopic data [357] to be at $T_{rev} \approx 190$ ps. The low temperature of the molecular beam, combined with its high stability and the extremely low pulse-to-pulse fluctuations of the laser system used, enabled for the first time the real-time detection of fractional revival structures in a molecular system.

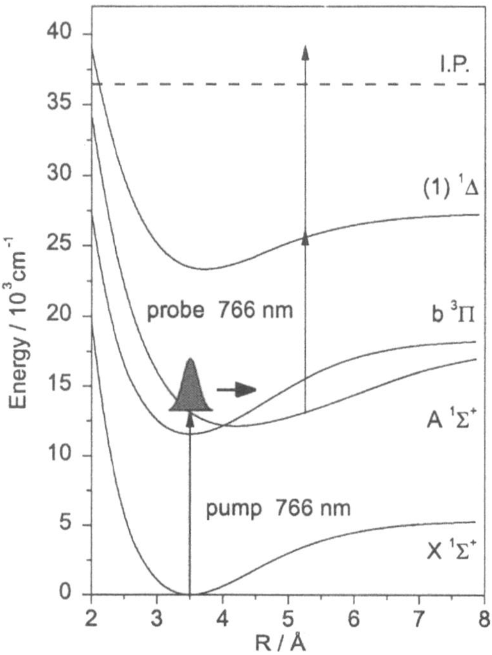

Fig. 3.32. Schematic illustration of the one-color pump&probe process in NaK (taken from [53]). The pump pulse prepares a wave packet at the inner turning point of the $A\,^1\Sigma^+$ state. The ionization geometry is located at the outer turning point of the molecule's vibration, where probably the $(1)\,^1\Delta$ state acts as a Franck–Condon window. The value for the ionization potential is taken from [358], the $(1)\,^1\Delta$ state potential curve is fitted to the values of [359], and the other curves are reproduced from the spectroscopic constants given in [360] (X state), [361] (b state), and [357] (A state)

The real-time spectra for ^{23}Na^{39}K and ^{23}Na^{41}K are illustrated in Fig. 3.31. In Fig. 3.31 a (^{23}Na^{39}K) a fast oscillation in the ion signal with a period $T_A^{(23,39)} \cong 441$ fs is clearly seen. The first maximum of this oscillation is found at ≈ 220 fs with respect to the temporal origin. The amplitude of the oscillation decreases within 20 ps. Around $T_{\rm frev}^{1,2(23,39)} \cong 75$ ps a pronounced recurrence of the oscillation is visible. A remarkable oscillation pattern appears for delay times around $T_{\rm frev}^{1,4(23,39)} \cong 40$ ps; there, a periodic variation of the ion signal with a period equal to $\frac{1}{2}T_A^{(23,39)}$ is clearly resolved.

The spectrum for ^{23}Na^{41}K (Fig. 3.31 b) is quite similar to the spectrum for the lighter isotope, although the signal-to-background ratio is not as high. This is due to the relatively small abundance of ^{41}K ($\approx$ 7%) compared with ^{39}K ($\approx$ 93%) in natural potassium. Here, the oscillation period $T_A^{(23,41)} \cong 446$ fs is slightly larger. The wave packet dephases within 15 ps. The first maximum is also measured at ≈ 220 fs. A time interval showing again the period $T_A^{(23,41)}$ can be recognized at approximately $T_{\rm frev}^{1,2(23,41)} \cong 68$ ps. However, this recurrence is not as pronounced as is the case of ^{23}Na^{39}K. Further oscillation patterns cannot be resolved directly from the real-time data.

The periods of 441 fs (^{23}Na^{39}K) and 446 fs (^{23}Na^{41}K) can be assigned to the classical oscillation period of a wave packet initialized by the pump pulse at $\Delta t = 0$. The vibrational levels v' of the $A\,^1\Sigma^+$ state in the range $8 \le v' \le 16$ are coherently excited. This observation is in good agreement with the period of 442 fs found by Berg et al. [353]. Since the first maximum occurs at about one half of the oscillation period T_A, one can assume the best location along the internuclear distance to ionize the excited molecules to be situated at the outer turning point of the A state's potential (see Fig. 3.32). This clearly demonstrates that only one Franck–Condon point exists in the two-photon ionization process. This finding is similar to the experimental and theoretical results for the one-color pump&probe experiments on K_2 at 834 to 840 nm and on Na_2 at 642 nm presented in Sects. 3.1.3 and 3.1.4 respectively. In the experiment described here, the $(1)^1\Delta$ state is the most probable candidate for a Franck–Condon transition because other electronic singlet states in the corresponding energy range lie rather low [359].

The dephasing time of ≈ 20 ps is fairly large compared to that measured by Berg et al. [353]. This might be due to the fact that in our experiment a beam of extremely cold molecules was used. Hence, only the lowest vibrational level of the electronic ground state is initially populated. In a heat-pipe oven, however, the molecules in general have higher vibrational temperatures. Starting from different initial levels leads to wave packet components which are prepared at slightly different internuclear distances on the PES of the A state. Therefore, the generated wave packet is not as localized as the wave packet prepared by pumping from one initial level only. Furthermore, the optical detection process used in our experiment takes place in a more defined geometry than in a detection scheme using collision-induced singlet–triplet transitions. Besides this, the occurrence of competitive relaxation processes starting from the intermediate $E\,^1\Sigma^+$ state in the scheme of Berg et al. might disturb the detected signal. Finally, the non-isotope-selective detection can easily influence the observed wave packet propagation (see Sect. 3.1.3).

To discuss the recurrences in the real-time spectra described above, the frequency components have first to be calculated by Fourier analysis. In Fig. 3.33, Fourier spectra of the real-time data for $^{23}Na^{39}K$ (Fig. 3.33 a) and $^{23}Na^{41}K$ (Fig. 3.33 b) are presented. Again the signal-to-background ratio for the heavier isotope is worse than for $^{23}Na^{39}K$, owing to the abundance ratio of ^{39}K and ^{41}K in natural potassium. Besides the frequency components corresponding to neighboring vibrational energy levels at $\omega_{A,1}^{(23,39)}$ and $\omega_{A,1}^{(23,41)}$, which have already been described by Berg and coworkers [353], here further frequencies are observed. Fourier components resulting from a coherent excitation of vibrational energy levels with $\Delta v' = 2, 3$ lead to the Fourier components $\omega_{A,2} \cong 150\,\mathrm{cm}^{-1}$ and $\omega_{A,3} \cong 225\,\mathrm{cm}^{-1}$, respectively, a phenomenon which is well known from experiments on I_2 [22, 23] and K_2 [43]. The difference between $\omega_{A,1}^{(23,39)} \cong 75.8\,\mathrm{cm}^{-1}$ and $\omega_{A,1}^{(23,41)} \cong 75.2\,\mathrm{cm}^{-1}$ manifests the isotopic shift of the two isotopes $^{23}Na^{39}K$ and $^{23}Na^{41}K$. No further frequencies originating from wave packet dynamics in other electronic states involved in the three-photon pump&probe scheme are observed.

The insets of Fig. 3.33 show the frequency $\omega_{A,1}$ in detail. The resolved frequency components are compared to the expected frequencies. The latter

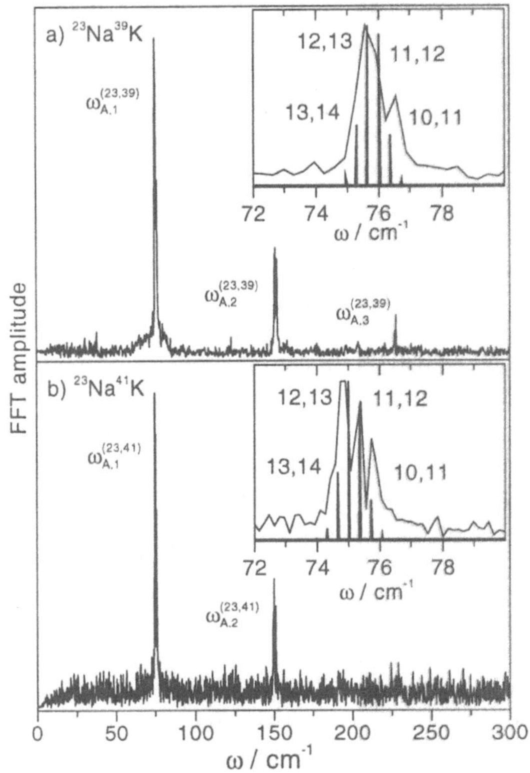

Fig. 3.33. Fourier spectra of real-time data for $^{23}Na^{39}K$ **(a)** and $^{23}Na^{41}K$ **(b)** (taken from [53]). Three frequency bands $\omega_{A,1}$, $\omega_{A,2}$ and $\omega_{A,3}$ are found. The insets show the first band $\omega_{A,1}$ greatly magnified and compared with frequencies of the vibrational spectrum calculated on the basis of data in [357]

are based on the spectroscopic constants for the $A\,^1\Sigma^+$ state calculated in [357] for an unperturbed A state. The excitation by the pump pulse with its spectral width of $190\,\mathrm{cm}^{-1}$ was taken into account. The vibrational energies for the heavier isotope were calculated using the isotopic correction factor

$$\varrho = \sqrt{\frac{\mu^{(23,39)}}{\mu^{(23,41)}}} \tag{3.33}$$

in the appropriate term of the polynomial expansion, where $\mu^{(\cdot,\cdot)}$ is the reduced mass of the respective isotope. The centers of the expected and the measured frequency distributions are in good agreement for both Fourier spectra. Owing to a reduced number of data points, the resolution of the frequencies in the real-time data is not as high as in the experiments performed on potassium dimers (Sect. 3.1.3). Therefore, a proper assignment to vibrational levels is difficult. This is especially not easy because some of the vibrational states involved may be perturbed by interaction with the $b\,^3\Pi$ state in the experimentally given rotational distribution. Since in Sect. 3.1.3 this was thoroughly investigated for K_2, a closer look at the revival structure of the real-time spectra is taken here.

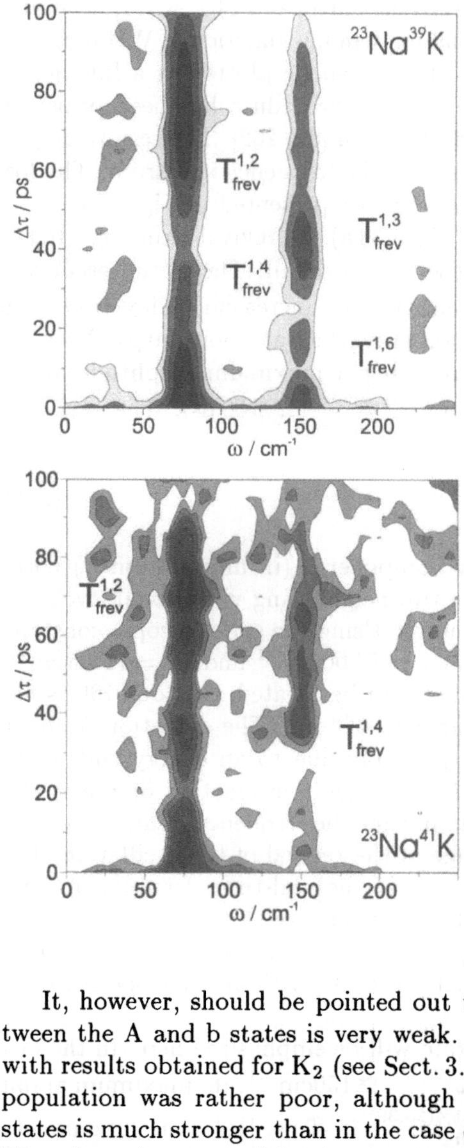

Fig. 3.34. Logarithmic contour plot of the spectrograms for $^{23}Na^{39}K$ **(a)** and $^{23}Na^{41}K$ **(b)** (taken from [53]). The Fourier amplitude increases from white to black. A half revival occurs at $T_{\mathrm{frev}}^{1,2}$. Further fractional revivals can be seen at $T_{\mathrm{frev}}^{1,4}$, $T_{\mathrm{frev}}^{1,6}$, and $T_{\mathrm{frev}}^{1,3}$

It, however, should be pointed out that the exchange of population between the A and b states is very weak. This can be deduced by comparison with results obtained for K_2 (see Sect. 3.1.3). There, the transfer of electronic population was rather poor, although the coupling between the A and b states is much stronger than in the case of the present NaK experiment [362]. Therefore, the norm of the A state population is nearly preserved (unlike, for instance the case of NaI as described by Engel [363]; see also [43]). The main influence of the crossing b state is the slight shift of the vibrational levels involved. This can, as demonstrated for K_2 [43], cause a significant change of the wave packet propagation compared with an unperturbed anharmonic PES. This, however, prevents a direct analysis as given by Vetchinkin and coworkers [364, 365] for an anharmonic system.

To discuss in detail the recurrences observed in the real-time spectra (Fig. 3.31) a spectrogram $I(\omega, \Delta t)$ has to be calculated for each isotope. For this purpose, as proposed by Vrakking, Villeneuve, and Stolow [51], a sliding-window Fourier transform with a Gaussian window function (FWHM = 2 ps) was applied. Then the Fourier amplitude I can be plotted as a function of the frequency ω and the delay time Δt; this procedure has been presented by Stolow and coworkers [51, 327] and by us in [52, 262]. The resulting spectrograms represent both the real-time and the frequency spectrum. The logarithmic contour plots of the spectrograms are presented in Fig. 3.34.

In the spectrogram for $^{23}Na^{39}K$ (Fig. 3.34 a) the revival structure already recognizable in Fig. 3.31 a can be observed in detail. The initial oscillation with $\omega_A^{(23,39)} \cong 75.8\,\mathrm{cm}^{-1}$ loses intensity with increasing delay time. The minimum amplitude for this oscillation is reached at about 40 ps. Then the oscillation becomes stronger again reaching a maximum amplitude at the recurrence time $T_{\mathrm{frev}}^{1,2(23,39)} \cong 75\,\mathrm{ps}$ ($\Rightarrow T_{\mathrm{rev}}^{(23,39)} \cong 150\,\mathrm{ps}$). The revival is expected to appear at a delay time

$$T_{\mathrm{rev}} = \frac{2}{c\,|\omega_1 - \omega_2|}, \tag{3.34}$$

where ω_1 and ω_2 are two frequency components (in units of cm^{-1}) corresponding to energy spacings between two neighboring vibrational levels and c is the speed of light in appropriate units. Using the spectroscopic constants given by Ross et al. [357], we obtain $\omega_1 = 76.00\,\mathrm{cm}^{-1}$ and $\omega_2 = 75.65\,\mathrm{cm}^{-1}$. Hence, the total and the half revival should be located at about 190 ps and 95 ps respectively, with respect to the zero of time. The deviation from the revival time measured here is most probably due to an energy shift of vibrational energy levels caused by the spin–orbit interaction of the $A\,^1\Sigma^+$ state with the $b\,^3\Pi$ state. This shift changes the frequency components and consequently leads to a drastic change of the revival of the oscillation. This nicely demonstrates the high sensitivity of the real-time detection method presented here. Nevertheless, the vibrational energy shifts seem to be small in comparison to those of the $A\,^1\Sigma_u^+$ state of $^{39,39}K_2$, where the structure of revivals consists of a pronounced beat structure with an envelope period of only 10 ps (Sect. 3.1.3).

A special feature in the spectrogram will be emphasized here. In the evolution of the frequency component $\omega_{A,2}^{(23,39)} \cong 150\mathrm{cm}^{-1}$, its maximum amplitude is reached at the delay time $T_{\mathrm{frev}}^{1,4(23,39)} \cong 40\,\mathrm{ps}$. Here, the fundamental frequency of the wave packet has its lowest amplitude. In the corresponding time interval in the real-time spectrum (Fig. 3.31 a) this frequency can also be seen directly. This revival-like structure for times smaller than the time for the total revival can be assigned to a fractional revival. The fractional revival found here at about half the time of the total revival can be understood semi-classically [54]: at the corresponding propagation time $T_{\mathrm{frev}}^{1,4}$ two neighboring frequency components, i.e. two frequency components originating from three vibrational energy levels, have a phase shift of π. The total wave packet then consists of partial wave packets oscillating against each other on the PES.

Therefore, in the experimental scheme (Fig. 3.32), ionization with the probe pulse is possible at two times within a classical oscillation period T_A. At $T_{frev}^{1,6} \approx 25$ ps and $T_{frev}^{1,3} \approx 50$ ps the corresponding fractional revivals consisting of the frequency $\omega_{A,3}^{(23,39)}$ occur in the spectrogram. Speaking classically, three frequency components oscillate at these delay times with phase shifts between each other of $\frac{2}{3}\pi$.

For the real-time data of $^{23}Na^{41}K$ (Fig. 3.34 b) a rather similar behavior in the spectrogram is seen. Here, the half revival is already seen at about 68 ps after the initial pump pulse. Following the ideas discussed in Sect. 3.1.3 for $^{39,39}K_2$ and $^{39,41}K_2$, the perturbation of some vibrational levels of the A state is expected to be stronger than in case of the lighter isotope. Although not directly seen in the real-time spectrum (Fig. 3.31 b), a fractional revival can clearly be identified around $T_{frev}^{1,4(23,41)} \cong 35$ ps in the spectrogram. Owing to the signal-to-background ratio in the real-time data caused by the abundances of the potassium isotopes, further fractional revivals cannot be resolved.

3.2 Ultrafast Wave Packet Propagation Phenomena in Excited Alkali Trimers

Vibronically coupled internal degrees of freedom are the main reason for the very distinct properties of small metal clusters. In diatomic systems, with their single vibrational mode, these phenomena do not yet occur. Therefore, an isolated oscillating dimer vibrates until eventually it radiates or predissociates. With an increasing number of participating atoms in a molecule, however the number of coupled vibrations drastically increases. This increase makes the identification of individual energetic redistribution channels extremely difficult, unless simple systems are chosen.

Owing to their relatively simple electronic structure, the alkali trimers can be regarded as such model systems, to study, for example the principles of intramolecular vibrational redistribution (IVR) in photoexcited molecules or clusters. Among the alkali trimers Na_3, especially when excited to its electronic B state, seems to be the best known. It acts in this section as a prototype for exploration of details of photoinduced IVR processes in real-time. Figure 3.35 sketches the principle of the experimental approach for triatomic s^1 systems such as the Jahn–Teller distorted Na_3 B system. Compared to the investigations on the dimer systems, the only difference is that the energy surfaces involved get a 'little' more complicated.

In Sects. 3.2.1–3.2.4 a detailed analysis of the sodium trimer excited to its electronic B state is presented. A short review of Na_3, with special interest in the spectroscopy of its electronic B state, is given in Sect. 3.2.1. Employing laser pulses of moderate intensities with durations of either 1.4 ps or 110 fs to excite the Na_3 B state, different vibrational modes of the excited trimer are detected selectively in the real-time spectra. While the picosecond laser

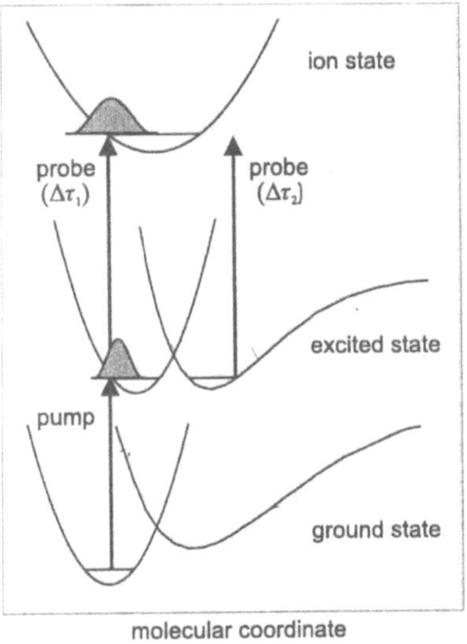

Fig. 3.35. Principle of real-time TPI spectroscopy of a triatomic s[1] system such as the Jahn–Teller distorted Na_3. A wave packet is prepared in an excited state of the neutral trimer by a pump pulse. Since in general the transition probability to the ion state is a function of the wave packet's location on the potential-energy surface, the propagation of the wave packet can be probed by a second, time-delayed ionization pulse. While for a delay time Δt_1 the ionization yield might be maximum, there is zero ionization probability for Δt_2

pulses yield preferential excitation of the slow pseudorotational mode with a period of 3 ps (Sect. 3.2.2), the use of $\sim$ 10 times shorter pulses allows the observation of the trimer's symmetric stretch mode with a 310 to 320 fs period for the first 5 picoseconds (Sect. 3.2.2).

These complementary experimental results can be explained to a great extent by quantum dynamical simulations of the real-time experiments. In Sect. 3.2.2, first the results obtained by means of two-dimensional (2d) ab initio potential-energy surfaces are briefly summarized. Even more sophisticated calculations are performed on three-dimensional (3d) ab initio potential-energy and transition dipole surfaces (Sect. 3.2.3). There, all three vibrational degrees of freedom of the Na_3 molecule are included in the theoretical treatment. The time-dependent wave packet dynamics elucidate the effect of ultrafast state preparation on the molecular dynamics. Extensive theoretical calculations indicate the possibility of initiating the molecular dynamics predominantly in selected modes during a certain time span by variation of the pump pulse duration (Sect. 3.2.4).

Another interesting triatomic candidate for femtosecond real-time spectroscopy is the potassium trimer. Although it has often been tried, conventional spectroscopy does not reveal any theoretically predicted electronic state of K_3. Ultrafast processes such as intermolecular vibrational redistribution (IVR) or even photodissociation on a picosecond timescale were considered responsible for this fact. With femtosecond real-time spectroscopy, however, the predicted electronic state (comparable to the Na_3 B state) should be detectable (Sect. 3.2.5).

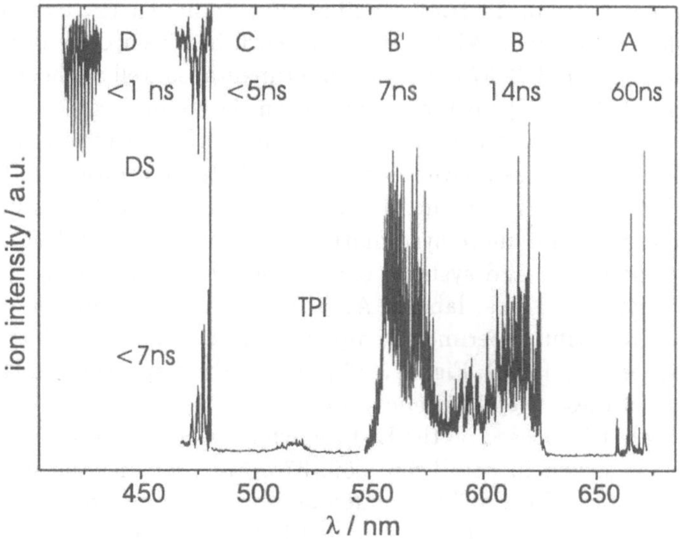

Fig. 3.36. Excited electronic states A, B, B′, C, and D of Na_3, revealing a pronounced substructure. While A, B, B′, and the energetically lower part of the C state were measured by resonant TPI; the D and parts of the C state were detected by depletion spectroscopy. The listed lifetimes were estimated by nanosecond pump&probe spectroscopy (taken from [369])

3.2.1 Brief Review of Spectroscopy of the Sodium Trimer

The alkali trimers show fascinating spectroscopic features, especially because they are Jahn–Teller systems revealing dramatic dynamical and optical effects. The electronic structure of these trimers M_3 is, because of their three valence electrons, particularly simple. In their ground state two of these electrons occupy a totally symmetric orbital, regardless of geometry [74, 117, 366, 367]. Consequently, the ground state ion M_3^+ is a compact equilateral triangle structure with D_{3h} symmetry. This structure is analogous to the H_3 molecule, for which essentially exact calculations are possible [368].

The equilibrium geometry of the neutral trimer, however, is not, as might at first be thought, an equilateral triangle. In the electronic ground (X) state, as well as in some of the excited electronic states, there is an unpaired electron in a degenerate electronic orbital [74]. This leads to a Jahn–Teller (or pseudo-Jahn–Teller) distortion away from the equilateral geometry towards an isosceles triangle geometry. In many of the electronic states there is only a small barrier between the various acute- and obtuse-angled isosceles triangle geometries, and the molecule changes from one geometry to another with relative ease. This is sometimes called ‘pseudorotation’.

The different structures of the neutral and ion ground states motivate a search for detailed information, especially about the intermediate states, i.e.

the excited electronic states of the trimer, and about internal energy transfer processes after photoexcitation. While very little work has been done onLi_3 [367, 370–373] and K_3 [367, 372, 373], a lot of experimental as well as theoretical investigations [72, 367, 373] have concentrated on the sodium trimer. The optical absorption spectrum of Na_3 was first experimentally observed by Hermann et al. [114] using mass-selective resonant two-photon ionization (TPI) spectroscopy. Delacrétaz et al. [68] and Broyer et al. [124, 374, 375] studied this absorption spectrum more systematically in the region of 1.75 eV to 3.75 eV. Five prominent band systems corresponding to transitions into different excited electronic states, labeled A, B, B′, C, and D, could be resolved by a series of beautiful experiments employing TPI as well as depletion spectroscopy (DS) (see e.g. [369]). Figure 3.36 presents these spectra and the estimated lifetimes of these five electronic states.

Here special interest is focused on the B state, which is bound. In Fig. 3.37 the TPI spectrum is shown in detail [68, 369, 376, 377]. Among the most important characteristics of the richly banded system is a long progression composed of nearly equally spaced bands ($\omega \approx 128\,\mathrm{cm}^{-1}$). This progression appears to be split into doublets. Besides this, a series of closely spaced bands fanning out from the doublet and increasing steadily in breadth accompanies each member of the main progression. A much weaker pattern of levels accounting for all remaining bands fit a harmonic series with $\omega \approx 137\,\mathrm{cm}^{-1}$. This spectrum has been explained in terms of the pseudorotational motion of the sodium trimer, which is now briefly explained.

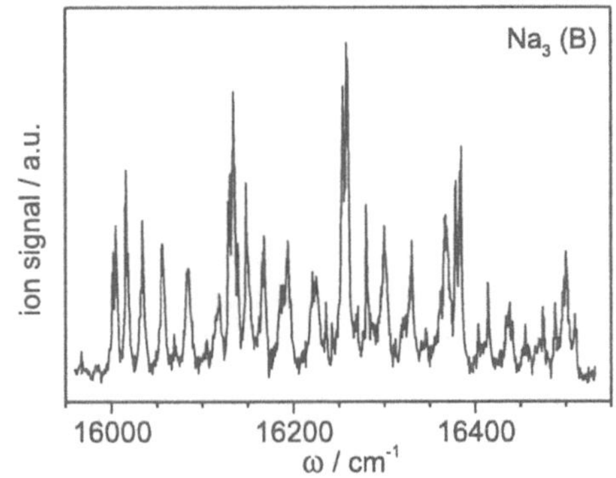

Fig. 3.37. Resonant TPI spectrum of the Na_3 B state (taken from [369])

The point symmetry group of the neutral sodium trimer is D_{3h}. In this symmetry group two of the three vibrational degrees of freedom (see Fig. 3.38), the symmetric bend Q_x and the asymmetric stretch Q_y, are energetically degenerate. The third normal coordinate is the totally symmetric stretch Q_s.

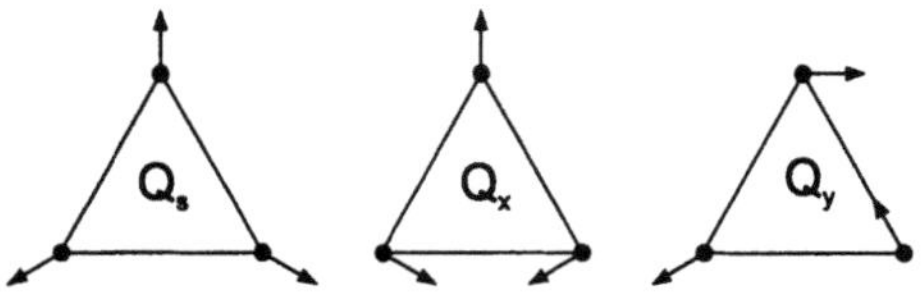

Fig. 3.38. Normal modes of Na_3(B). Q_s, Q_x and Q_y correspond to the symmetric stretch, symmetric bend, and asymmetric stretch mode respectively

The superposition of the degenerate Q_x and Q_y modes leads to a pseudorotational motion of the molecule (see Fig. 3.39), which may also be described in a polar coordinate system using the coordinates ϱ and φ defined by

$$\varrho = \sqrt{Q_x^2 + Q_y^2} \quad \text{and} \quad \varphi = \arctan \frac{Q_y}{Q_x}. \tag{3.35}$$

The angle φ is referred to as the angular pseudorotation coordinate, and the distance ϱ as the radial coordinate.

The B state's interesting features can be regarded as an exciting challenge for real-time spectroscopy. First it is of great interest whether it is possible to observe the time-resolved pseudorotation (see Sect. 3.2.2). Next, the question arises of why the breathing mode Q_s is nearly absent in cw or nanosecond TPI spectra. At first glance, there seems to be no argument as to why the Q_s mode could not be excited. Ultrafast internal vibrational redistribution might cause the immediate disappearance of this mode in cw and nanosecond TPI spectra. However, femtosecond real-time spectroscopy might open a temporal window to study the vibronic coupling of the Na_3 B system in its initial phase and perhaps prepare selectively this specific vibrational mode (see Sect. 3.2.4).

3.2.2 Real-Time Spectroscopy of Na_3 B

The transition of the excited electronic B state of Na_3 is located between 600 nm and 625 nm. In order to perform real-time two-photon ionization spectroscopy in this spectral region the setup shown and described in Fig. 2.2 is used (see Sect. 2.1.1). It provides laser pulses of moderate intensities with durations of either ~ 1.4 ps or ~ 120 fs to excite the Na_3 molecule to its electronic B state and subsequently to ionize the excited molecule. The bandwidth of the picosecond pulse is 40 cm^{-1}, while for the femtosecond pulse 260 cm^{-1} is estimated. Therefore, in both cases several vibrational states of the Na_3 system can be excited simultaneously Figure 3.40 illustrates schematically the pump&probe cycles.

Real-Time Pseudorotation of the Sodium Trimer. A typical real-time spectrum obtained with picosecond excitation is shown in Fig. 3.41. The excitation wavelength is 620 nm, which corresponds to an intense resonant peak in the TPI spectrum. A clear beat strucure, symmetrical about the zero of time, is observed. The period amounts to ~ 3 ps. The oscillation is damped with a time constant of about 3.5 ps. A constant offset is present, due to TPI

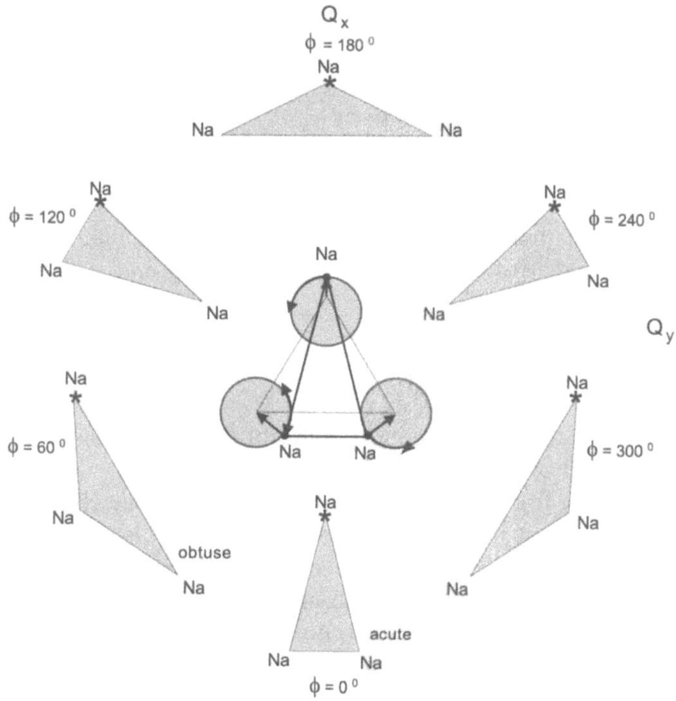

Fig. 3.39. Pseudorotation of the Na_3 (B) molecule : a Cartesian space in Q_x (symmetric bend mode) and Q_y (asymmetric stretch mode) is sketched. At the origin of this space the sodium trimer has equilateral geometry. Simultaneous vibration in Q_x and Q_y causes a rotation of each of the three atoms on a circle of radius ϱ around their position in the equilateral geometry. For different angles φ which the atoms pass through on their way around the circle, the molecule has different geometries. At $\varphi = 0°, 120°, 240°$ the trimer has an obtuse isosceles triangle geometry, whilst for $\varphi = 60°, 180°, 300°$ the shape is an acute isosceles triangle. The continuous change of the molecular configuration looks at first glance to be a rotation of the molecule, but is not as the marking (*) of the upper sodium atom directly shows. Hence this motion of the molecule is sometimes called pseudorotation

processes induced by either a single pump or a single probe pulse. There is very little difference found in the oscillation pattern on changing slightly the excitation wavelength.

The first theoretical calculations to describe these real-time spectra were based on a simple two-dimensional model of the vibrating/pseudorotating sodium trimer [380]. Ab initio energy surfaces served as guidance for constructing model surfaces. The molecular dynamics of the Na_3 B system were simulated by the representative time-dependent wave packet $\Psi_B(Q_x, Q_y, t)$. Its norm

$$I(t) = |\Psi_B(Q_x, Q_y, t)|^2 \tag{3.36}$$

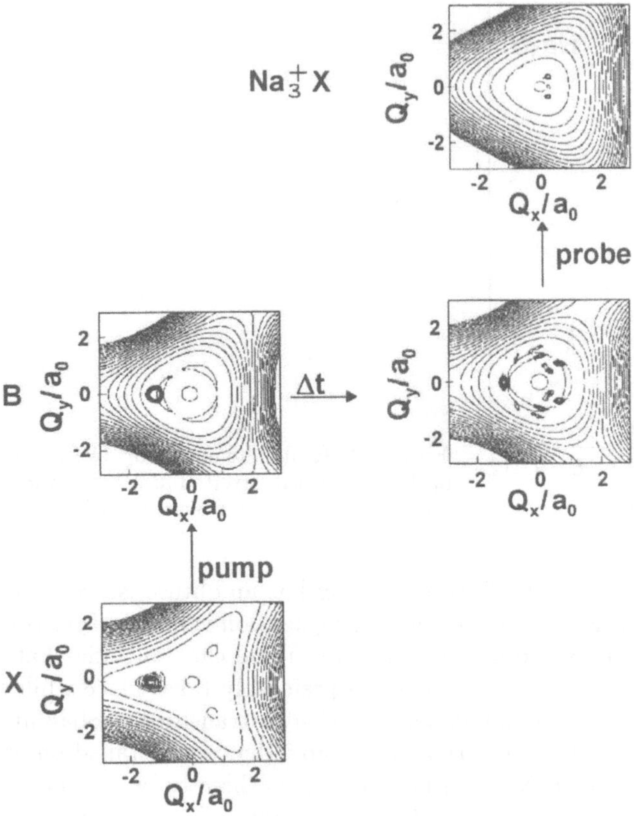

Fig. 3.40. Pump&probe scheme for Na_3 B state: The sequence of snapshots shows the initial wave function $\Psi_X(0)$, and the excited wave functions $\Psi_B(t)$ and $\Psi_B(t+\Delta t)$ at the times of the maxima of the pump and probe pulses (for the representative delay time 3 ps) and $\Psi_{X^+}(3\,\mathrm{ps})$, superimposed on the ab initio potential-energy surface of Na_3 X, Na_3 B, and Na_3^+ X, respectively. The 2d contour plot shows the favorable Franck-Condon windows of the pump and probe laser pulses for obtuse geometries of Na_3 (taken from [378])

nicely corresponds to the 3 ps oscillation of the real-time spectrum. The oscillation is assigned to the pseudorotation of the excited trimer.

Figure 3.42 presents, for several delay times, snapshots of the wave packet propagation represented by the autocorrelation function. After a vertical X → B transition the wave packet, generated on the B state surface at one of its three local minima, starts propagating. After ~ 1.5 ps the wave packet is localized at the three equivalent saddle points of the PES, corresponding to a minimum of the ion yield in the real-time spectrum. The saddle point can be identified with the acute isosceles triangle geometry (see Fig. 3.43). For a delay time of ~ 3 ps the wave packet is found concentrated at the three equivalent minima of the B state surface. This is the situation where

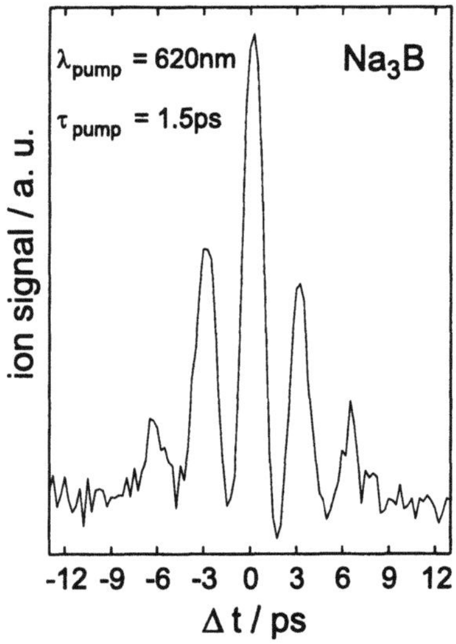

Fig. 3.41. Real-time one-color TPI spectrum of the Na_3 B state excited at $\lambda = 620$ nm with laser pulses of 1.4 ps (taken from [379])

maximum ion yield is found. There the trimer has an obtuse isosceles triangle shape (Fig. 3.43). Further wave packet propagation is now a repetition of many of these sequences. From this comparison one can conclude that the transition from the B state to the ion state preferably takes place while the trimer has an obtuse triangular shape. However, a transition probability of nearly zero is found for the acute triangular shape. Hence, the transition pathway of the TPI process in Na_3 can be determined directly by this analysis. The result of this investigation provides the starting point for more realistic three-dimensional model simulations of the pseudorotating Na_3 B state (see Sects. 3.2.3 and 3.2.4).

Real-Time Observation of the Sodium Trimer's Breathing Mode. The experiment discussed above was now repeated under similar conditions, but the pump pulse width was reduced to ~ 120 fs. The normalized one-color real-time spectrum is presented in Fig. 3.44 for delay times $-3\,\mathrm{ps} < \Delta t < 8\,\mathrm{ps}$ between the pump and probe pulses. A clear oscillation in the ion signal's temporal evolution, symmetrical about the zero of time is dominant. The modulation of the ion signal amounts 20 % of the total averaged ion signal. The origin of the symmetry is due to the fact that at $\Delta t = 0$ the roles of pump and probe pulses exchange. The oscillation period $T_S \approx 320$ fs can be directly observed in the real-time spectrum. The amplitude of the oscillation loses its initial height within 12 periods. At delay times greater than 6 ps the oscillation is not present any more.

In Fig. 3.45 the Fourier spectrum is shown. A single frequency component at $104\,\mathrm{cm}^{-1}$ is clearly visible. This frequency can be directly correlated with the 320 fs oscillation found in the real-time spectrum and can be assigned to the symmetric stretch mode Q_s of the excited Na_3 molecule. Surprisingly, no further Fourier components are seen in the spectrum. By comparison with the

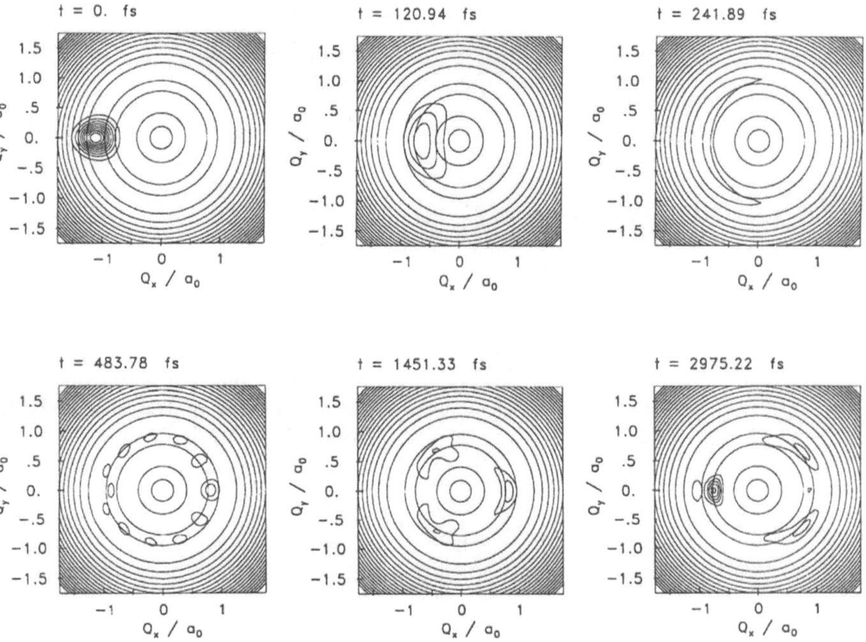

Fig. 3.42. Vibrational dynamics of Na_3 B, illustrated by snapshots of the norm of the representative wave packet on model potential-energy surfaces for different delay times (taken from [81])

cw TPI spectra of the Na_3 B state [68] one would expect a strong frequency component at $128\,\mathrm{cm}^{-1}$ corresponding to the radial part of the trimer's pseudorotation and additional low-frequency components representing the angular movement of the pseudorotation.

In previous femtosecond pump&probe experiments [131] performed with pulse durations of 60 fs and high laser pulse powers, eleven frequency components could be estimated. There, besides the Q_s mode of the B state, the asymmetric stretch Q_y and bending Q_x frequencies of the ground state appeared with intensities in the same order of magnitude. These two frequencies can be explained by excitation by resonant impulsive stimulated Raman scattering (RISRS). Additionally, the Q_x and Q_y modes and also the radial and angular components of the pseudorotation were identified in [131]. However, owing to the right choice of the laser pulse parameters in the real-time experiment shown here, a single mode instead of a mixture of different molecular vibrations was excited in the present experiment. This enables femtosecond state preparation, and as a consequence this mechanism might in future be used to efficiently control mode-selective subsequent reactions. Besides this, the coupling and the internal vibrational redistribution can be analyzed on a femtosecond timescale. Here, especially, the spectrogram technique is the right tool. In Fig. 3.46 the ultrafast vibrational redistribution is visible as a

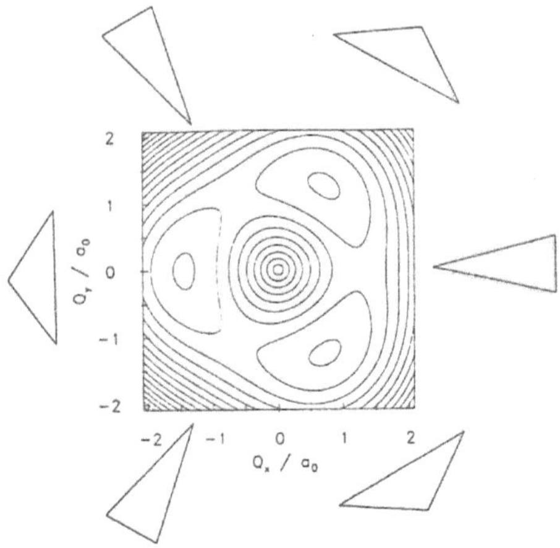

Fig. 3.43. Contour plot of the Na_3 B energy surface with the corresponding obtuse and acute isosceles triangle geometries of the trimer (taken from [380])

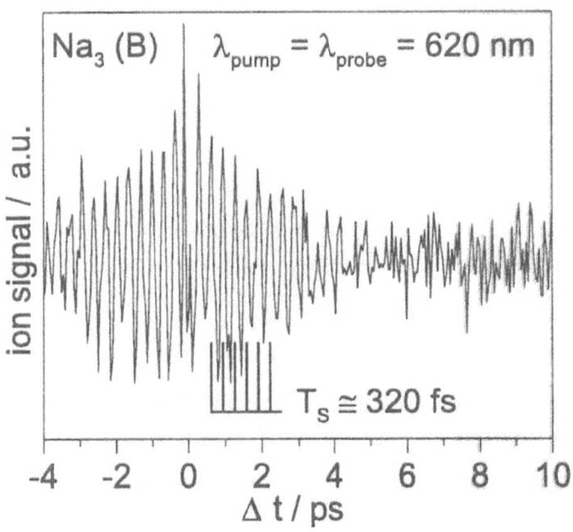

Fig. 3.44. Real-time spectrum of the Na_3 B state. The pump pulse parameters are pulse width $\sim$ 120 fs (FWHM) and spectral width 10.1 nm (taken from [381])

decrease of the Q_s mode within ~ 6 ps. This result is in excellent agreement with the simulations of the wave packet dynamics presented in Sect. 3.2.4, as well as those in [62, 81, 295, 382, 383].

3.2.3 3d Approach to the Wave Packet Dynamics of the Na_3 B State

The quantum calculations used to describe the wave packet dynamics of Na_3 are carried out in three dimensions here. These three dimensions correspond to the three normal modes of an equilateral triangle, which is formed by the Na_3 molecule in D_{3h} symmetry (see Fig. 3.38).

The simulation of a real-time spectrum requires the solution of the time-dependent Schrödinger equation for a system of n potential-energy surfaces which are coupled by a laser field:

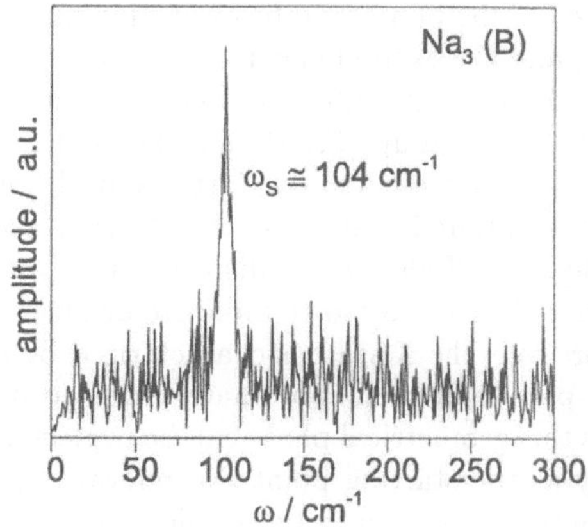

Fig. 3.45. Fourier spectrum of the Na_3 B state. The pump pulse parameters are pulse width $\sim$ 120 fs (FWHM) and spectral width 10.1 nm (taken from [381])

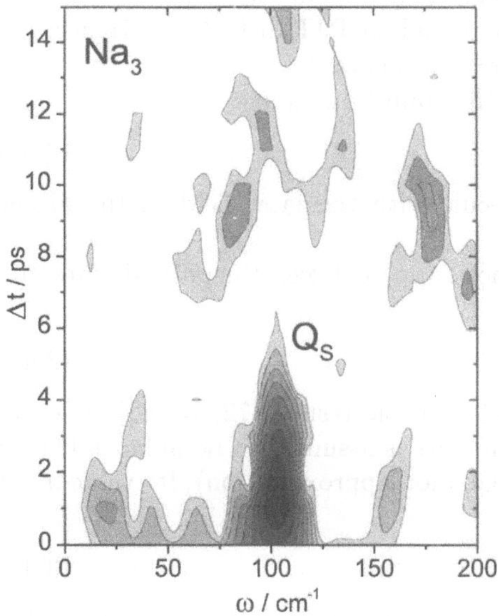

Fig. 3.46. Spectrogram of the Na_3 B state excited with $\sim$ 120 fs (FWHM) laser pulses (taken from [381])

$$\mathrm{i}\hbar\frac{\partial}{\partial t}\begin{pmatrix}\Psi_{\mathrm{X}}\\ \Psi_{\mathrm{B}}\\ \Psi_{\mathrm{I}(E')}\end{pmatrix} = \begin{pmatrix} H_{\mathrm{XX}} & H_{\mathrm{XB}} & 0\\ H_{\mathrm{BX}} & H_{\mathrm{BB}} & H_{\mathrm{BI}(E')}\\ 0 & H_{\mathrm{I}(E')\mathrm{B}} & H_{\mathrm{I}(E')\mathrm{I}(E')}\end{pmatrix}\cdot\begin{pmatrix}\Psi_{\mathrm{X}}\\ \Psi_{\mathrm{B}}\\ \Psi_{\mathrm{I}(E')}\end{pmatrix}. \tag{3.37}$$

In the case of the sodium trimer the Hamilton matrix of the corresponding Schrödinger equation takes a 3×3 form. The diagonal elements H_{ii},

$$H_{ii} = T + V_i \quad \text{with} \quad i = \mathrm{X}, \mathrm{B}, \mathrm{I}, \tag{3.38}$$

of the unperturbed Hamiltonian describe the kinetic energies T_i plus the potential energies V_i for the three states involved.

The neutral adiabatic PES involved in the pump&probe excitation mechanism, the electronic ground state X, and the excited electronic B state (see Fig. 3.48) were evaluated by a three-dimensional interpolation of ab initio results calculated by the Bonačić-Koutecký group [72, 73, 82]. For the ion state Na_3^+ I, a slightly modified analytical fit presented by Carter and Meyer [384, 385] was adopted. The electronic ground state and the B state are subject to Jahn–Teller (JT) or pseudo-Jahn–Teller (PJT) distortions respectively. Hence, the ideal equilateral triangle is distorted to an acute or obtuse triangle which is lower in energy. Moreover, the X state is characterized by a geometrical phase along the angular pseudorotation coordinate φ due to its JT effect [76, 386]. In the simulations this geometrical phase can be neglected because the ground state serves only as the starting point. No relevant dynamics take place on this PES. Furthermore, as seen in the studies presented here, PJT couplings of the B state to the higher PESs do not play any important role and therefore may be neglected. In contrast to the two neutral states, the ion state does not exhibit any JT or PJT distortions. Hence, the equilateral geometry is lowest in energy in this case.

The off-diagonal elements H_{ij} of the Hamiltonian in

$$H_{ij}(t) = -\mu_{ij} E(t) \tag{3.39}$$

describe the interaction of the molecule with the laser field in the dipole approximation.

The dipole transition moment function μ_{ij} between the ground state and the B state

$$\mu_{\mathrm{BX}} = |\langle \varphi_{\mathrm{el,B}} | eQ | \varphi_{\mathrm{el,X}} \rangle| \tag{3.40}$$

has been obtained by interpolation of ab initio results [72, 73, 82]. The unknown μ_{IB} between the B and the ion state is assumed to be independent of the nuclear coordinates (generalized Condon approximation). Its value is set to

$$\mu_{\mathrm{IB}} = 3ea_0, \tag{3.41}$$

which is similar to the value of μ_{XB}. As long as both pulses do not overlap in time, the time-dependent electromagnetic field $E(t)$ is given by the sum of the pump and the probe pulse

$$E(t) = E_{\mathrm{pump}}(t) + E_{\mathrm{probe}}(t). \tag{3.42}$$

The electromagnetic field of a single pulse (pump or probe) can be described classically by

$$E_{\mathrm{pump/probe}}(t) = E_0 \cos(\omega t) s(t). \tag{3.43}$$

E_0 represents the amplitude of the electromagnetic field, ω the laser frequency, and $s(t)$ the shape function of the laser pulse. Here, a Gaussian shape function approximates the experiments rather well [62, 81, 306, 379]. The functions Ψ_i ($i =$X, B, I(E')) denote the time-dependent wave packets on the relevant PES. Their temporal evolution in space is evaluated by a Fast Fourier transform (FFT) propagation [299, 301]. In practice, the wave packets are represented on Cartesian spatial and temporal grids:

$$Q_\mathrm{x} = Q_\mathrm{y} = -2.9a_0 \text{ to } 2.9a_0 \quad \text{with} \quad \Delta Q_\mathrm{x} = \Delta Q_\mathrm{y} = 0.092a_0, \tag{3.44}$$

$$Q_\mathrm{s} = 6.3a_0 \text{ to } 8.501a_0 \quad \text{with} \quad \Delta Q_\mathrm{s} = 0.071a_0, \tag{3.45}$$

$$t_k = k\Delta t \quad \text{with} \quad \Delta t = \frac{2\hbar}{E_\hbar} = 4.838 \cdot 10^{-17}\,\mathrm{s}. \tag{3.46}$$

The time-independent vibrational eigenvalues E_v and eigenfunctions $\xi_{\mathrm{X_v}}$ of the X state are obtained by solving the time-independent Schrödinger equation

$$H_{\mathrm{XX}}\xi_{\mathrm{X_v}} = E_v\xi_{\mathrm{X_v}} \tag{3.47}$$

for the electronic ground state X by means of the sinc-function discrete variable representation (DVR) technique [297].

Since both experiments (picosecond and femtosecond) were carried out with low temperatures in the molecular beam of approximately 50 K, the initial vibrational state is assumed to be a coherent linear superposition of the three lowest delocalized vibrational eigenfunctions $\xi_{\mathrm{X_{v=0-2}}}$ of the X state, which are very close in energy [62, 75, 79, 81, 380, 387]. Therefore, the Boltzmann distribution yields nearly identical populations of these three eigenfunctions at the relevant temperature. The linear superposition leads to an initial wave packet that is localized in one of the equivalent minima of the X state:

$$\begin{pmatrix} \Psi_\mathrm{X}(q, t=0) \\ \Psi_\mathrm{B}(q, t=0) \\ \Psi_\mathrm{I}(q, t=0) \end{pmatrix} = \begin{pmatrix} \sum_{\mathrm{v}=0}^{2} \xi_{\mathrm{X_v}}(q) \\ 0 \\ 0 \end{pmatrix}. \tag{3.48}$$

The rotational temperature of 10 K in the molecular beam leads to rotational quantum numbers in the range of 10 to 20. Using the rotational constants given by Ernst and Rakowsky [70], rotational periods are estimated to be several hundreds of picoseconds. Consequently, rotational recurrences, which might interfere with the observed vibrational beat structure, appear for the first time after more than 1 ns. Therefore, the influence of the rotational motion can be neglected in the simulation, as we concentrate on the dynamics up to a few picoseconds.

There is an additional complication, due to the fact that the probe laser pulse produces Na_3^+ ions and electrons. The electron which is ejected can have a continuum of kinetic energies E'. Different methods have been developed to simulate this continuum [302, 303, 304, 305]. In the simulations the method used for the femtosecond real-time spectroscopy of the K_2 molecule (Sect. 3.1.3 and e.g. [351]) was tested. Detailed investigations showed that under certain conditions the contribution of a single, optimally selected kinetic energy E' already defines the main features of the ion signal and allows a careful analysis of the transition pathway. This method works especially well if the range of the kinetic energies E' is quite small. In the case of the sodium trimer the probe laser pulse has just enough energy to reach the minimum of the ion state. Hence, the relevant energy range covers only $1.3 \times 10^{-3} E_{\mathrm{H}}$ ($E_{\mathrm{H}} = 4.3597482 \times 10^{-18}$ J: Hartree energy) and the kinetic energy of most of the electrons is nearly zero. This assumption is also confirmed by the experimental ZEKE (electrons with zero kinetic energy) signal which is quite similar to the ion signal [71, 131].

The theoretical pump&probe signal was simulated by carrying out many propagations in which the delay time between the pump and probe pulses was varied systematically. At the end of such a propagation the population of the ion state was calculated for a given delay time. Many of these propagations, with different delay times, then yielded the full theoretical pump&probe signal. As one can imagine, in the case of a three-dimensional problem like that of Na_3 such a simulation is very time-consuming. Therefore, besides the 'exact quantum dynamical' (QD) method an approximate method to calculate the pump&probe ion signal was also used. This method is based on the following steps.

- Some full propagations with different delay times between the pump and the probe laser pulse in order to elucidate the dominant transition pathway.
- Determination of the relevant excitation regions of the B state, which allow the molecule to be excited into the ion state.
- The start of a propagation, considering the X and B states to be coupled by the pump laser pulse only.
- Calculation of the norm of the wave packet in the relevant excitation region (ER) of the B state.

Using this method, the value of

$$I_{\mathrm{ER}}(t) = \langle \Psi(Q,t) | \Psi(Q,t) \rangle_{Q_{\mathrm{ER}}} \tag{3.49}$$

then yields the approximate pump&probe signal.

The calculated norm $I_{\mathrm{ER}}(t)$ in the relevant excitation regions reflects the dynamics taking place on the PES investigated, as only the parts of the wave packet in the relevant excitation region can actually be excited into the ion state in a real pump&probe experiment. The advantage of this method is that instead of many propagations on the three PESs involved, just one sufficiently long propagation is performed on two surfaces coupled by the pump pulse. By means of this approximate method, it is possible to test different free parameters of the laser pulse in a reasonable time. However, this method cannot be used in every case. For instance it will fail in the case of an unrestricted relevant excitation region into the ion continuum. Multiphoton processes induced by high-intensity laser pulses mean that a single excitation pathway no longer dominates and these situations can also not be analyzed well by this method. In the following we refer to this method as the 'approximate' (QD) method, while inclusion of all three PESs as well as the pump and probe pulses will be called the 'exact' (QD) method.

In this context 'exact' (QD) means the most exact ab initio treatment to describe the vibrational dynamics during MPI processes induced by ultrafast laser pulses which is possible and practicable with the current state-of-the-art computational methods. As already pointed out in detail, this method takes into account all three vibrational modes and only neglects the rotational degrees of freedom. The 'exact' (QD) method can describe all laser-induced MPI processes without exception and is able to record the main features of the vibronical dynamics without the use of any fitting parameter.

In the case of the sodium trimer (see Sect. 3.2.4) the relevant excitation regions in the B state have been determined as between $7.0a_0$ and $7.2a_0$ in the Q_s coordinate; the centers of the excitation regions in the pseudorotation coordinates lie at $\varphi = 60°, 180°, 300°$ and $\varrho = 0.56a_0$. These regions have a diameter of $0.24a_0$.

3.2.4 Pulse-Width-Controlled Molecular Dynamics

Most of the pump&control experiments carried out so far have used diatomic molecules, because in such simple systems, with only one vibrational degree of freedom, the dynamics can be controlled relatively easily. In larger molecular systems with three or more vibrational degrees of freedom, the situation becomes much more complicated and it is an interesting question whether the concept of 'controlled molecular dynamics' can still be realized. Here, it is shown that different vibrational modes of the sodium trimer can be selectively excited during an electronic excitation with ultrashort laser pulses. For this reason, it should in future be possible to control subsequent reactions. The relevant control parameter in these investigations is the duration of the pump pulse.

For this, various 3d quantum ab initio simulations of the wave packet dynamics in Na_3 B are presented here and compared to ultrashort laser pump&probe experiments. In addition to 'exact QD' calculations, an 'approximate QD' method is suggested to simulate the main features of a pump&probe spectrum. The simulations provide satisfactory results in comparison to 'exact QD' calculations. By means of these two methods it is possible to reproduce and to explain the different experimental pump&probe spectra. The 310 fs oscillation in the femtosecond pump&probe experiment [62, 81] can clearly be assigned to the Q_s vibration, while the 3 ps oscillation of the picosecond pump&probe experiment [306, 379] is caused by a slow pseudorotational wave packet motion.

Real-Time Spectrum Obtained with 120 fs Pump Pulses. In order to simulate the femtosecond experimental ion signal, the following pulse parameters were used: duration $\Delta t_{\mathrm{FWHM}} = 120\,\mathrm{fs}$, intensity $I = 520\,\mathrm{MWcm}^{-2}$

and central wavenumber $\bar{\nu} = 0.073E_{\mathrm{H}} = 16\,021\,\mathrm{cm}^{-1}$. The result of the femtosecond pump&probe experiment is shown in Fig. 3.47. The upper curve corresponds to the experimental real-time signal, the curve in the middle shows the 'exact QD' calculated signal and the lower curve the 'approximation'. All three curves clearly exhibit a dominant oscillation with a period of approximately 310 to 320 fs. As shown in earlier 'exact QD' studies [62, 81], the 120 fs pulse excitation drives the excited wave packet along the Q_{s} coordinate and this oscillation period is reflected in the ion signal. During the first 4 to 5 ps, the symmetric stretch is the dominantly excited vibration of the molecule when a 120 fs laser pulse is used. At longer times, an energy transfer from the symmetric stretch to both pseudorotational vibrations takes place [62, 81]. The agreement between the 'exact QD' simulation and the experimental result shown in Fig. 3.47 is excellent.

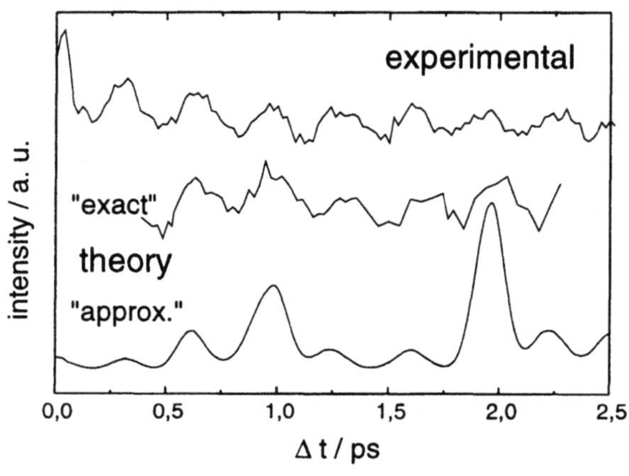

Fig. 3.47. Comparison of experimental (*upper curve*), 'exact QD' theoretical (*middle curve*) and 'approximate QD' theoretical (*lower curve*) real-time spectra of Na_3 B for a 120 fs excitation at $\bar{\nu} = 16\,021\,\mathrm{cm}^{-1}$ (taken from [382])

The 'approximate QD' pump&probe signal (Fig. 3.47) still yields a good agreement with the experimental and the 'exact QD' theoretical results. The 320 fs oscillation structure dominates in all cases. Moreover, one of the slower pseudorotational vibrations with a period of about 1 ps can also be detected by our 'approximate QD' method. A detailed analysis by means of the Fourier transform of the corresponding autocorrelation function [378] and of the induced wave packet dynamics reveals that this oscillation is caused by a slow pseudorotational vibration in the coordinate φ. As the Fourier spectrum shows the wave packet prepared in the B state is centered around 621 nm, i.e. between the vibrational states $\nu_{\varphi} = 5$ and 6 of the φ mode. The energy distance between these two eigenstates corresponds to a vibrational period of about 1 ps. It is one of the fastest pseudorotational vibrations which can be observed in the absorption spectrum and is the next slower vibration after the Q_{s} mode. This interpretation of the 1 ps oscillation is confirmed by analyzing the induced wave packet dynamics. Here, an accumulation of the wave packet at $\varphi = 0°$ and $180°$ after half a period (500 fs) can be observed. This oscillation is also obvious in the 'exact QD' pump&probe signal, but not as dominant. Evidently the probe laser pulse reduces the effect of this vibration in the pump&probe signal, which might be due to the fact that the details of the excitation region are not determined exactly unlike the 'exact QD' simulation.

Summarizing the strengths of the 'approximate QD' method, it reflects the dominant wave packet motions quite well, while the intensities cannot be reproduced exactly. Nevertheless, the comparison between the 'exact QD' and the 'approximate QD' method demonstrates that the latter method enables an efficient study of the dynamics that are induced with different laser pulse parameters even for larger molecules.

Real-Time Spectrum Obtained with 1.5 ps Pump Pulses. The 'approximate QD' method was applied to simulate the picosecond pump&probe experiment. An 'exact QD' simulation of an experiment with laser pulses of only a few picoseconds duration is hardly possible with the computers available today. The following free parameters of the laser pulse were adopted for the picosecond experiment: duration $\Delta t_{\mathrm{FWHM}} = 1.5\,\mathrm{ps}$, intensity $I = 300\,\mathrm{MW\,cm^{-2}}$, and central wavenumber $\tilde{\nu} = 0.073 E_{\mathrm{H}} = 16\,021\,\mathrm{cm^{-1}}$. The dynamics induced by such a laser pulse are illustrated by means of representative snapshots in Fig. 3.48. At the bottom of these snapshots the ground state X and at the top the excited state B are shown in pseudorotational coordinates Q_{x}, Q_{y} for a Q_{s} value of $7.15 a_0$. The first snapshot illustrates the situation when the pump laser pulse reaches its maximum intensity. This time is used as the temporal origin. Parts of the ground state wave packet have been transferred in a Franck–Condon-like transition to the B state. Here, the wave packet presents the obtuse geometry of the sodium trimer. As already pointed out, the wave packet can be excited into the ion state from this position. Therefore, the theoretical 'approximate QD' real-time signal exhibits a maximum (see lower curve in Fig. 3.49). The second snapshot of Fig. 3.48 reflects the situation after 1.5 ps. The wave packet has moved to the other side of the trough and represents the acute geometry. From this area a transition into the ion state cannot take place, which is reflected by a minimum in the theoretical pump&probe curve. The third snapshot (Fig. 3.48) shows the wave packet after 3 ps. Now it has turned back close to its original position, where it again represents the obtuse shape. Hence, the pump&probe signal exhibits a second maximum.

Figure 3.49 shows a comparison between the experimental (top) and the 'approximate QD' pump&probe spectrum (bottom). Both curves show an oscillation with a period of about 3 ps, which can be identified with the vibration from the obtuse to the acute geometry as described above (including only $v = 1$ and 2, with an approximate energy spacing of $13\,\mathrm{cm^{-1}}$ of the φ mode). This is in accordance with an earlier study [380] where the 3 ps oscillation was explained by a similar pseudorotational motion of the molecule,

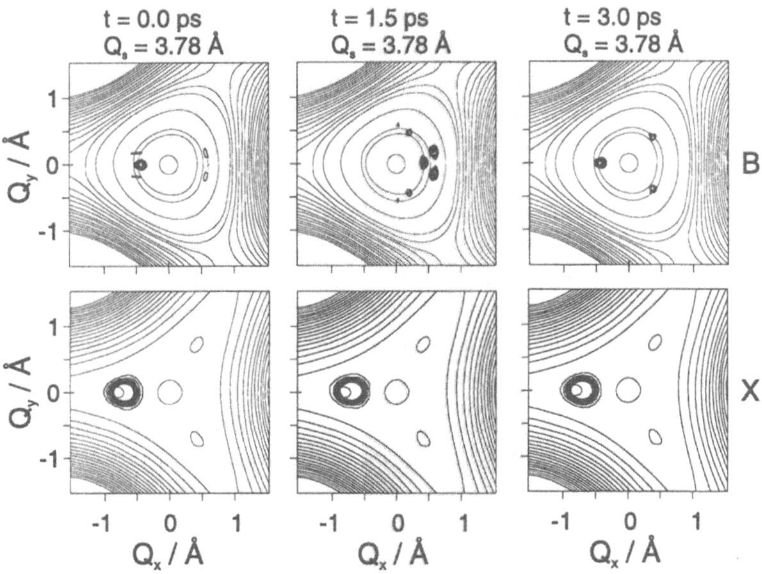

Fig. 3.48. Snapshots during the picosecond pump&probe excitation of Na_3 B for selected delay times between pump and probe pulses (taken from [382]). *Bottom*, X state and *top*, B state, shown in the pseudorotational coordinates Q_x, Q_y for a Q_s value of $7.2a_0$

in a two-dimensional model. The full three-dimensional simulations including the pump laser pulse confirm this result, so that the origin of the beat structure is fully clarified.

Another question that cannot be answered yet – even in the three-dimensional simulation – and thus remains open concerns the decay of the experimental real-time spectra on a timescale of 6 ps. Several possible explanations exist as to why this decay cannot be observed in the simulation. One reason could be the existence of a dark state (e.g. a quartet state) that crosses the B state and which is not considered in the present simulations. In this case the wave packet could be transferred to the dark state by an intersystem crossing process. However, as far as we are aware, the ab initio calculations do not yield any crossing state in the energy region of the B state [73]. Hence, this explanation is not supported by the calculations.

However, an explanation in the form of a hypothesis can be given here. The decay observed in the experimental signal originates from the low pulse intensities applied and basically reflects the overlap of the pump and probe laser pulses. The decay does not occur in the calculated ion signal, because the probe laser pulse is neglected in the present simulations. As is seen in the comparison of the 'exact QD' and 'approximate QD' results (Fig. 3.47), the probe laser pulse can influence the intensities of the spectrum. In the case of picosecond excitation the laser pulses are rather long. The total pulse duration can be approximated as 6.8 ps, when the intensity has decreased to

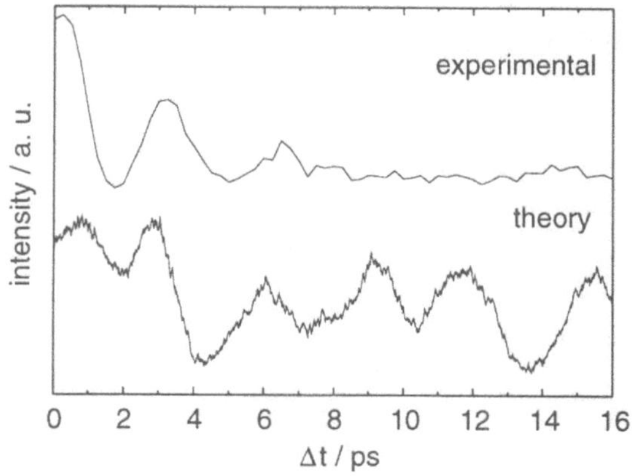

Fig. 3.49. Comparison of experimental (*upper*) and 'approximate QD' theoretical (*lower*) real-time spectra of Na_3 B for a 1.5 ps excitation at $\bar{\nu} = 16\,021\,\mathrm{cm}^{-1}$ (taken from [382])

$0.0025 I_{\max}$. During this time the pump and probe pulses can overlap. When both pulses overlap, the total intensity increases which leads to a higher transition rate into the ion state. With increasing delay time between pump and probe pulses the total intensity of course decreases. Consequently, the total ion signal is reduced, which is nicely reflected by the decay of the experimental signal. Owing to the moderate laser intensities ($520\,\mathrm{MW\,cm^{-2}}$) chosen in the experiment, there is enough intensity to excite the molecule into the ion state only, while the pump and probe pulses overlap. In the experimental real-time spectra, the last, already very weak peak can be observed at a delay time of 6 ps. Here, both pulses still slightly overlap. For longer times, when both pulses do not overlap any more, the intensity of the probe laser pulse is too small to excite the molecule into the ion state and so the signal vanishes. Thus the decay of the ion signal corresponds to the overlap time of the pump and the probe pulses.

This hypothesis might be confirmed by 'exact QD' pump&probe simulations, which are – as already mentioned – very time-consuming at present, or by repeating the experiment with higher laser intensities. Then the 3 ps oscillation should continue after the initial decay for longer.

Influence of the Pulse Duration on the Dynamics. Besides the simulation and interpretation of the experimental results described here, a theoretical investigation was carried out in which the influence of the pulse duration on the induced wave packet dynamics was extensively tested. For this purpose the 'approximate QD' method was applied. In these calculations the duration of the pump pulse was varied systematically. In the first series, different femtosecond laser pulses were tested. The results are shown in Fig. 3.50 a. The upper curve represents the extreme and hypothetical case of a 0 fs laser pulse, which means that the wave packet is excited instantaneously to the B state. In the middle the signal for the 120 fs excitation is shown again, and at the bottom the resulting curve of a 200 fs pulse is presented. Comparing

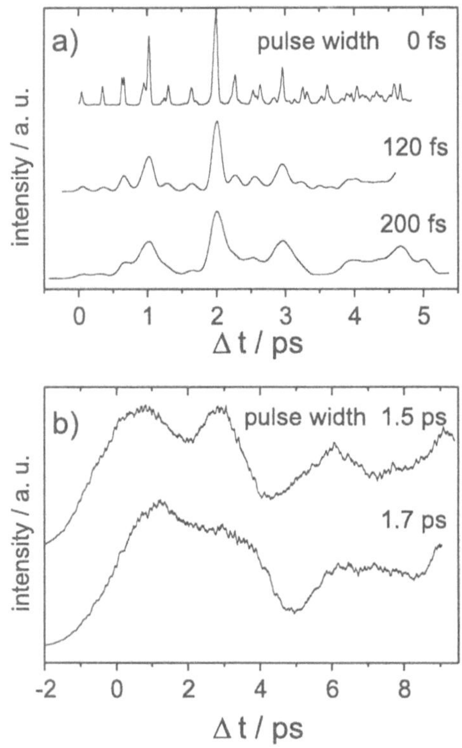

Fig. 3.50. Theoretical simulation of the variation of the pump pulse width (taken from [382]). **(a)** Results for variations in the femtosecond time domain, **(b)** results for the picosecond time domain

these three curves demonstrates nicely that the 'resolution' of the observed dynamics decreases with increasing pulse duration.

As already discussed, the 'approximate QD' signal of a 120 fs pulse exhibits two dominant vibrations which are caused by the Q_s mode (310 to 320 fs) on the one hand, and by a slow pseudorotational mode (1 ps) on the other hand. Interestingly, the vibration along the Q_s mode loses its pronounced character after 3 ps (see Fig. 3.51), i.e. the wave packet spreads out along this coordinate, as a movie of the three-dimensional wave packet dynamics of the B state shows [388]. This observation has already been interpreted as an effect of the anharmonicity of the PES and as intramolecular vibrational redistribution from dominantly Q_s to ϱ and φ vibration [62, 81]. The behavior described is also reflected by the 'approximate QD' pump&probe signal where the oscillation of period 310 to 320 fs caused by the Q_s mode vanishes after a time of 4 to 5 ps (see also Fig. 3.52).

For the hypothetical case of a 0 fs pulse a similar behavior of the 'approximate QD' pump&probe signal concerning the Q_s and the slow pseudorotational φ mode can be observed (upper curve in Fig. 3.50a). But, in contrast to the 120 fs case, a radial vibration along ϱ with a period of 260 fs is now also detected. In the first and second peaks at 50 fs and 340 fs, both modes are still unresolved owing to their similar vibrational frequencies. The

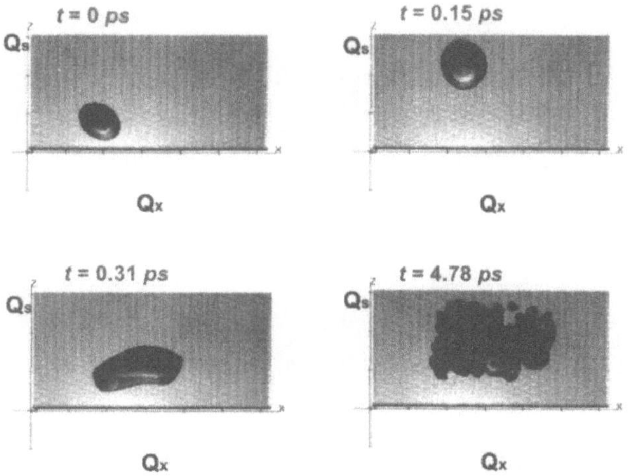

Fig. 3.51. Snapshots of wave packet propagation in Na_3 B for a 120 fs excitation at $\tilde{\nu} = 16\,021\,\mathrm{cm}^{-1}$ demonstrating the dephasing of the 310 fs breathing mode (taken from [382])

first splitting between them occurs in the third peak, at 630 fs, and can be observed for the next peaks too. A movie of the corresponding wave packet dynamics confirms this analysis [382]. The vibrations along both the radial and the Q_s coordinate can be detected clearly during the first 3 ps. Looking at the curves for the 0 fs and for the 120 fs pulse one could conclude that in the 120 fs case the radial mode is also excited, but cannot be detected owing to the inferior 'resolution'. That this is not the case has been shown in [62, 81], (see Fig. 3.47), where the starting wave packets along ϱ in both cases are compared. While the center of mass of the wave packet induced by a 0 fs pulse is clearly shifted with respect to the equilibrium distance, allowing clear oscillations in this mode, in the 120 fs case the starting wave packet is nearly located in the potential minimum. This is the result of the different energy widths and Franck–Condon factors for the different pulse durations.

The lowest curve in Fig. 3.50a, representing a pulse width of 200 fs, shows that the slow pseudorotational vibration with a period of 1 ps dominates. The Q_s mode can hardly be recognized. In this case the pulse duration is already too long to resolve the faster ϱ or Q_s vibrations, but it is still short enough to detect the slower pseudorotational mode.

A similar behavior can also be observed for the picosecond laser pulses (Fig. 3.50b). While a pulse width of 1.5 ps is able to resolve the 3 ps pseudorotational vibration (centered at $\nu_{Q_s} = 1$ and 2), this is no longer possible with a pulse width of 1.7 ps. In this case a slower pseudorotational motion with a period of about 8 to 9 ps begins to dominate. This vibration is also present using a pulse width of 1.5 ps, as a comparison with Fig. 3.49 (lower curve) shows.

Summarizing the results described above, it has been confirmed that the induced wave packet dynamics depend strongly on the pulse duration. This is not surprising in so far as the pulse width has to be clearly shorter than the vibrational motion which is to be detected. However, it is interesting that

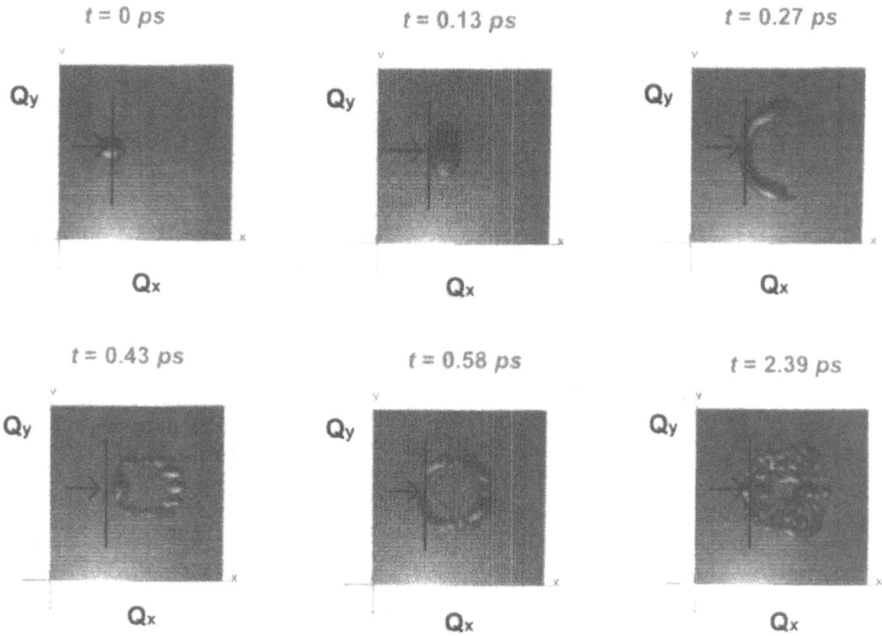

Fig. 3.52. Snapshots of wave packet propagation in Na_3 B for a 120 fs excitation at $\bar{\nu} = 16\,021\,\mathrm{cm}^{-1}$ showing the transfer from the breathing mode to the pseudorotation mode (taken from [382])

though ultrashort laser pulses with a broad energy span are used, different vibrational modes can be excited quite selectively. Of course, the prepared wave packets are coherent superpositions of eigenstates excited by the laser pulse. But owing to the pulse duration and, of course, favorable Franck–Condon factors, several eigenstates can dominate the wave packet (see for instance the 120 fs pulse where the competing ϱ mode almost vanishes in comparison to the Q_s mode). This demonstrates that it is possible to prepare dominantly selected vibrations even in larger molecules. This leads to the possibility of carrying out laser-controlled reactions.

Furthermore, in combination with the experimental real-time studies the theoretical three-dimensional investigations demonstrate nicely that femtosecond spectroscopy is a powerful tool to achieve the dominant excitation of different vibrational modes by a proper choice of the laser parameters. This phenomenon has been known within the community for a long time as a working hypothesis. Here, documentation and proof of this effect is given.

The possibility of preparing specific vibrational modes by ultrashort pulses provides an understanding of how the oscillation is built up and finds its pseudorotational rhythm, which has also been observed in the cw absorption spectrum [68]. This allows one to control mode-selective subsequent reactions [389] in a time domain shorter than the competing IVR processes. For instance, in the case of the Q_s mode excitation, a subsequent reaction has to be induced within 4 ps before this vibration loses its dominant character. Thus the possibility of controlling reactions has now also been demonstrated for a triatomic molecule. It will be stimulating to extend the present investigations to larger systems to determine, if this kind of mode sensitivity is conserved.

3.2.5 The Dissociative K_3 Molecule

A further interesting and promising candidate is the K_3 molecule. Theoretical calculations [82] have predicted an electronic state at about 800 nm, comparable to the Na_3 B state. However, to the best of my knowledge this electronic state has never been observed, although various highly sensitive methods such as multiphoton ionization and depletion spectroscopy, using cw or pulsed nanosecond laser sources were used [83]. Hence, exciting the potassium trimer in this spectral region with ultrashort pulses might allow the observation of both the wave packet propagation, as seen in the case of the Na_3 B system, and the ultrafast dissociation processes in real time. This would provide a deeper insight into the nature of the interaction leading to energy transfer between the bound states involved and of the repulsive potential energy surfaces. Apart from this, the potassium trimer case represents a great challenge for femtosecond spectroscopy to demonstrate its unique experimental power in obtaining information on the dynamics, otherwise unattainable.

For several excitation wavelengths the femtosecond real-time dynamics of the trimer has been observed. In Fig. 3.53, a representative time evolution of the 3PI signal for excitation of K_3 with 70 fs laser pulses at a central wavelength of 798 nm is shown. In the center, which marks the zero-of-time, the interferometric autocorrelation peak is clearly observable (see also Fig. 2.23). The transient ion signal is symmetric with respect to this peak as is expected for a one-color 3PI experiment. A fast decay is clearly seen and is superimposed on a 450 fs oscillation. Two processes are involved, wave packet propagation on the PES and ultrafast dissociation. Hence, in the following discussion the ion signals are analyzed in two steps. First the ultrafast decay is discussed, followed by an analysis of the superimposed coherent oscillation.

Ultrafast Photodissociation. Since radiative decay of the trimer is in the time domain of nanoseconds, this population can be changed either by photodissociation or by intersystem crossing to an electronic state, from which, under these experimental conditions, no ionization can take place. However, in the latter case this behavior should be directly visible as a drastic modulation of the observed wave packet propagation [43, 329, 315]. But this is not found. Apart from this, the first theoretical calculations of the PES of K_3 gave no evidence of the existence of any 'dark state' in this energy regime. Therefore, it seems to be reasonable that ultrafast photodissociation causes the fast decay of the ion signal. The K_2 can also be ionized by the probe pulse,

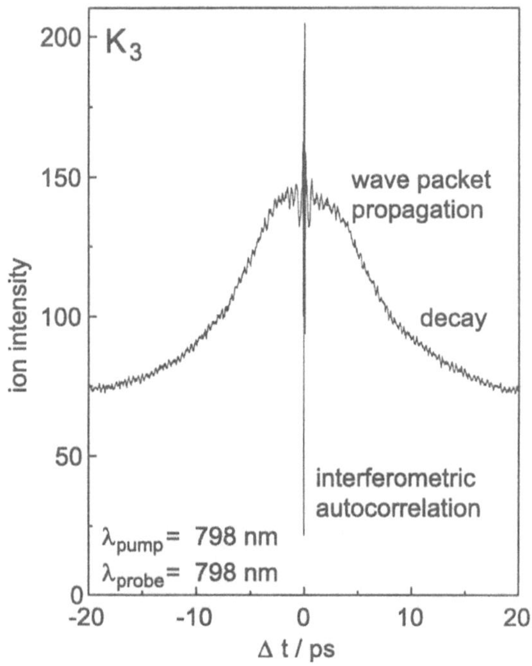

Fig. 3.53. Real-time evolution of the 3PI signal for K_3 excited at $\lambda = 798\,\text{nm}$ (taken from [260]). The symmetric shape is due to the applied pump and probe pulses of similar wavelength. Therefore, at the zero of time the pulses interchange their roles. Besides the ultrafast decay a superimposed oscillation with $T = 450\,\text{fs}$ is visible. Around the zero of time the interferometric autocorrelation as shown in detail in Fig. 2.23 is seen

resulting in a weak increase of the K_2 signal. This increase, however, is so small compared to the total amount of K_2 in the beam, that demonstration of the presence of these fragments is not possible.

To describe the decay of the real-time spectra, one might think of using the simple energy-level model described in Sect. 2.2.2. However, the experimental data (ignoring the fast oscillation) do not at all fit to a convolution of the overall system response with a single exponential decay. Therefore, the extended energy-level model developed in Sect. 2.2.2 was applied. The real-time spectra were fitted with the convolution function:

$$f_{\text{ext}}(t) = \left\{ N_0 \exp\left(-\frac{t}{\tau}\right) + \alpha \left[\exp\left(-\frac{t}{\tau_1}\right) - \exp\left(-\frac{t}{\tau_2}\right)\right]\right\} * \ell(t), \quad (3.50)$$

with the constant α depending on τ_1, τ_2, and the number of initially excited larger clusters. The decay times τ_1, τ_2 are due to fragmentation of larger clusters $K_{n>3}$ into K_3 and re-fragmentation of these products respectively. $\ell(t)$ is the overall system response of the measuring system, being represented by the envelope of the interferometric autocorrelation of the laser pulses. The value of interest is $1/\tau$, which can be identified with the photodissociation probability of the excited potassium trimer. For further details see Sect. 2.2.2.

The function $f_{\text{ext}}(t)$ was fitted to the experimental data $I(t)$ by means of a least-squares fit routine. In Fig. 3.54, for different wavelengths λ of the pump pulse, the resulting fits are shown as lines in comparison to the measured data, and are in excellent agreement. The dependence of the decay time τ

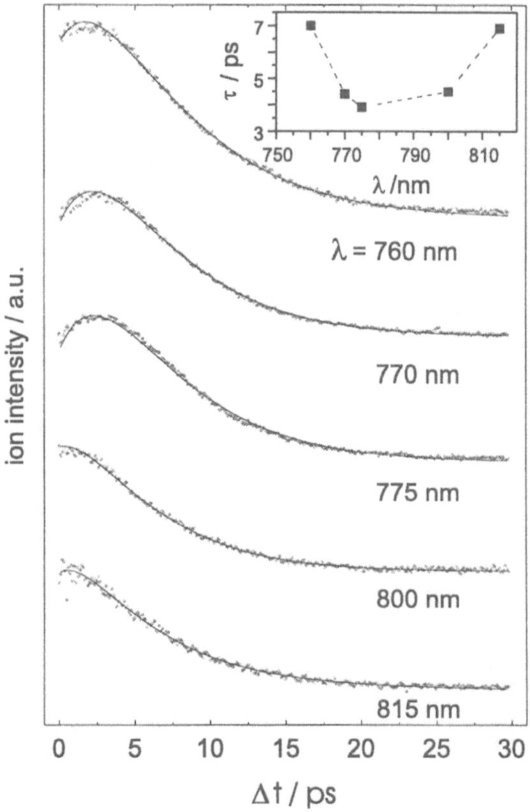

Fig. 3.54. Temporal evolution of the 3PI signal of K_3 for five different wavelengths of the pump pulse compared with the least-squares fit function $f_{\text{ext}}(t) = n(t) * \ell(t)$ ($\cdots$ experimental data; — fit function). Inset: dependence of the photodissociation lifetime τ on the wavelength λ of the pump pulse. The data were obtained by a least-squares fit procedure (taken from [260])

on the wavelength of the pump pulse was estimated, as depicted in the inset of Fig. 3.54. The values of τ found vary between 4 and 7 ps. A minimum is found for $\lambda = 790\,\text{nm}$ with $\tau = 4\,\text{ps}$. The reason might be that in the wavelength domain examined, several electronic states with different lifetimes in the range 4 to 7 ps exist. Here, detailed theoretical calculations of the K_3 potential-energy surfaces are required.

In summary, a fast photodissociation process coupled with a rather low oscillator strength might be the reason why cw spectroscopy was unable to detect this electronic state. Femtosecond spectroscopy with high peak power and broad spectral width of the exciting laser pulses, combined with the probing of the excited electronic state within a few picoseconds, however, opens a time window to efficiently detect this state and its dynamics.

Wave Packet Propagation. After applying a femtosecond pump pulse, a wave packet will be prepared in the excited state of K_3. This wave packet will propagate on the PES. The real-time evolution of the ion signal (Fig. 3.53) clearly reveals the wave packet propagation. However, the wave packet loses intensity with each oscillation period. This can be explained by the dissocia-

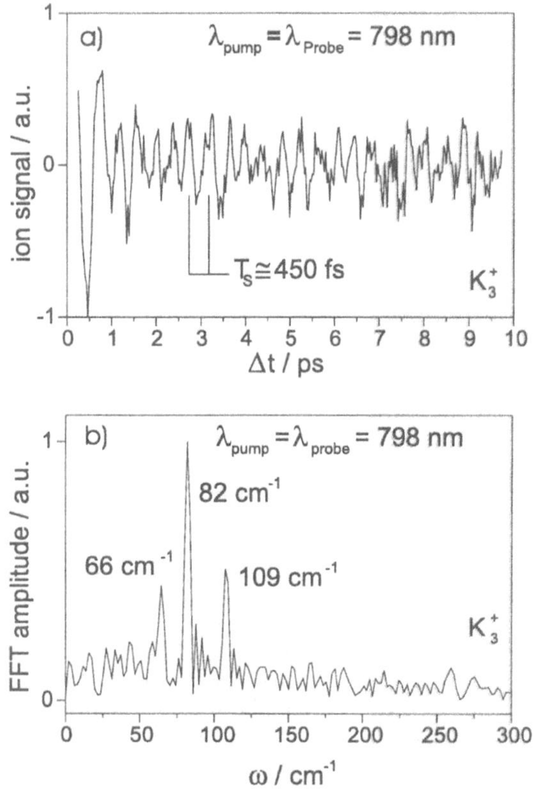

Fig. 3.55. (**a**) Real-time evolution of the 3PI signal for K_3 excited at $\lambda = 798\,\text{nm}$ (taken from [260]). The first 10 ps, with a clear oscillation of period $\sim 450\,\text{fs}$, are shown. The original data have been deconvoluted to overcome the decay. (**b**) Fourier spectra of the 3PI signal of K_3 for an excitation wavelength $\lambda = 798\,\text{nm}$. Three dominant lines are visible. They are assigned to three normal modes of the trimer.

tive character of this excited state, which is visible in the overall decay of the ion signal. With higher resolution, Fig. 3.55 a presents this oscillation with its dominant period of $T \sim 450\,\text{fs}$.

Since this oscillation is not at all a pure 450 fs vibration, a mixture of several modes of the trimer might cause the observed oscillation pattern. For a detailed analysis it is, therefore, necessary to perform a Fourier analysis of the transient spectra. In Fig. 3.55 b the corresponding Fourier spectrum is depicted. It reveals three frequencies with wavenumbers $\bar{\nu}_1 = 66\,\text{cm}^{-1}$, $\bar{\nu}_2 = 82\,\text{cm}^{-1}$, and $\bar{\nu}_3 = 109\,\text{cm}^{-1}$. In Table 3.3 the corresponding vibrational periods and intensities are listed.

By comparison with data for the ground (X) state [390] and excited electronic B state [29, 74, 131] of Na_3, these values can be assigned to normal modes of the trimer's ground and excited states. Since, as a first approximation,

$$\bar{\nu} \propto \sqrt{1/m}, \tag{3.51}$$

with m the mass of a single atom of the trimer, a ratio

$$\kappa = \bar{\nu}_{K_3}/\bar{\nu}_{Na_3} \tag{3.52}$$

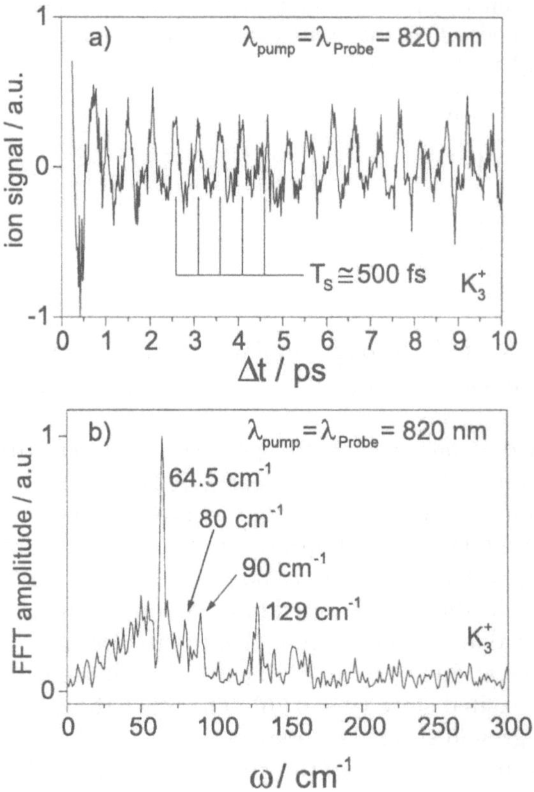

Fig. 3.56. (**a**) Real-time evolution of the 3PI signal for K_3 excited at $\lambda = 820\,\text{nm}$ (Taken from [381]). The first 10 ps, with a clear oscillation of period ~ 500 fs, are shown. The original data have been deconvoluted to overcome the decay. (**b**) Fourier spectra of the 3PI signal of K_3 for an excitation wavelength $\lambda = 820\,\text{nm}$. Two dominant and two weak lines are visible. They are assigned to normal modes of the trimer.

can be introduced to compare the vibrational modes of the alkali trimers. With $m_K = 39$ amu and $m_{Na} = 22.9$ amu one obtains $\kappa = 0.78$.

The dominant line at $82\,\text{cm}^{-1}$ can be assigned to the symmetric stretch-mode Q_s^* of the excited system. Comparing this frequency with the data in [29, 74, 131], where $\bar{\nu}(Q_s^{Na_3\,B}) = 105\,\text{cm}^{-1}$ was found, the ratio is $\kappa = 0.78$. This value is in excellent agreement with our estimate, demonstrating its usefullness. However, the two weaker lines at $66\,\text{cm}^{-1}$ and $109\,\text{cm}^{-1}$ have no corresponding known vibrational frequencies of Na_3 excited to its bound B state. Since the peak intensities of the applied 70 fs pulses, $\geq 1\,\text{GW}\,\text{cm}^{-2}$, are rather high it seems that both lines document the ground state dynamics of the trimer. Similarly to the results of Gerber [29, 131] for Na_3 and de Vivie-Riedle et al. [328] for K_2 using resonant impulsive stimulated Raman scattering (see [328] and references therein), a wave packet is generated in the ground state. These dynamics are also reflected in the real-time one-color 3PI signal. By comparison with the vibrational frequencies of the Na_3 X system [390], the line at $66\,\text{cm}^{-1}$ can be assigned to the asymmetric bending mode Q_y^X of the ground state. With $\bar{\nu}(Q_y^{Na_3\,X}) = 87\,\text{cm}^{-1}$ from [390], the value of κ is 0.76, which again is in reasonable agreement with our approximate

Table 3.3. Wavenumber $\bar{\nu}$, vibrational period T, and relative intensity I (compared to Q_s^*) of the symmetric stretch mode Q_s^* of the excited state and the asymmetric (Q_y^X) and symmetric stretch (Q_s^X) modes of the K_3 ground state

K_3 modes	$\bar{\nu}$ / cm^{-1}	T / fs	I /a.u.
Q_y^X	66	505	0.45
Q_s^*	82	406	1
Q_s^X	109	306	0.52

value $\kappa = 0.78$. The line at $109\,\text{cm}^{-1}$ belongs similarly to the symmetric stretch mode Q_s^X of K_3. Comparison with $\bar{\nu}(Q_s^{Na_3\,X}) = 140\,\text{cm}^{-1}$, [390], gives $\kappa = 0.78$.

Similar results are obtained if one excites the potassium trimer with $\lambda_{\text{pump}} = 820\,\text{nm}$. An oscillation with a period of nearly 500 fs is now observable, again superimposed on a rapid decay. Figure 3.56 presents the real-time (Fig. 3.56a) and the Fourier (Fig. 3.56b) spectrum. Four distinct frequencies with wavenumbers $\bar{\nu}_1 = 64.5\,\text{cm}^{-1}$, $\bar{\nu}_2 = 80\,\text{cm}^{-1}$, $\bar{\nu}_3 = 90\,\text{cm}^{-1}$ and $\bar{\nu}_4 = 129\,\text{cm}^{-1}$ are visible in the Fourier spectrum. Again, comparing these values with the known values of the sodium trimer enables quite a good assignment. The weak line at $80\,\text{cm}^{-1}$ can be identified with the symmetric stretch mode Q_s^* of the excited state ($\kappa = 0.76$). The other three lines seem to belong to the trimer's ground state. The frequency $64.5\,\text{cm}^{-1}$ can be ascribed to the asymmetric bending mode Q_y^X of the ground state ($\kappa = 0.74$). The line at $90\,\text{cm}^{-1}$ cannot be assigned with certainty. By comparison with the experimental data of Baumert et al. [132] and the theoretical work of Thompson et al. [373], it might reflect the symmetric stretch mode of the trimer's ground state. The fourth line, at $129\,\text{cm}^{-1}$, can be interpreted as the second harmonic of the Q_s^* mode ($64.5\,\text{cm}^{-1}$). This line is due to coherent excitation of vibrational levels with a difference in the corresponding quantum numbers of $\Delta v = 2$.

The spectrogram technique allows the visualization of the modes' interplay on a femtosecond timescale. Figure 3.57 neatly reveals the fast relaxation of the broad band of excited modes in less than 5 ps into one favored mode of the trimer. This example once more nicely demonstrates the strength of the spectrogram method as a tool in modern femtosecond spectroscopy.

As performed for the sodium trimer, configuration interaction ab initio calculations of the PES, combined with time-dependent quantum dynamical simulations, are essential for further comparisons and for a detailed picture of the dynamics.

Owing to the fast photodissociation in K_3, femtosecond spectroscopy with its excitation pulses of high fluence and temporal resolution provided the only possible choice to find experimentally this theoretically predicted electronic state, and cw spectroscopy failed. Apart from this, femtosecond spectroscopy

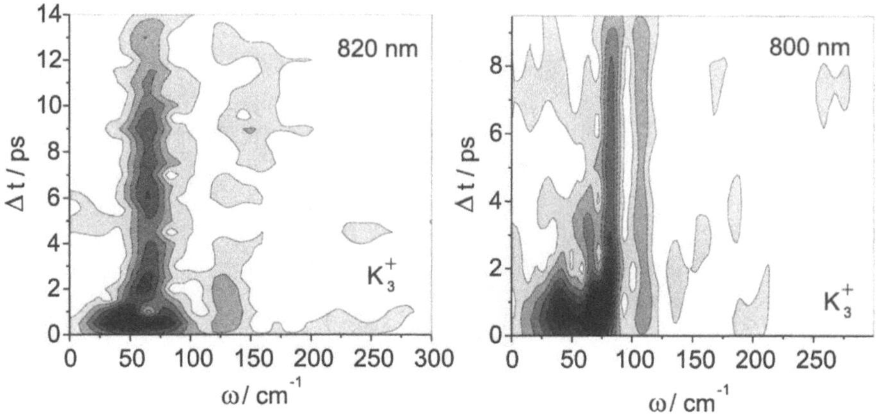

Fig. 3.57. Spectrograms of the 3PI signal for K_3 excited at $\lambda = 820\,\mathrm{nm}$ and $\lambda = 800\,\mathrm{nm}$ [391]. The sliding Gaussian window of the Fourier transformation has a width of 1 ps (FWHM)

allows a study of the entire ultrafast vibrational and dissociative dynamics of the state.

In comparison to the Na_3 system, the K_3 molecule exhibits outstanding features. While the Na_3 B state (Sects. 3.2.2, 3.2.4 and [62, 71, 378–382]) reveals wave packet propagation but no dissociation, and the neighboring C state (Sect. 4.1 and [392]) shows only fast photodissociation, the K_3 system presents both wave packet phenomena and ultrafast photodissociation simultaneously. The K_3 molecule, therefore, can be regarded as a limiting case or model system. Theoretical investigations similar to those of the Na_3 B state might give detailed new insight into the intramolecular processes of wave packet propagation in K_3.

4. Ultrafast Photodissociation

In Chap. 3, wave packet propagation could be observed for nearly all of the alkali dimer and trimer systems considered, over a rather long time compared to the wave packet oscillation period. The wave packet dynamics – a fingerprint of the excited molecule – definitely characterize the excited bound electronic state of these molecules. However, with the results on K_3 (excited with $\lambda \sim 800\,\text{nm}$), another phenomenon, which often governs ultrafast molecular and cluster dynamics, comes into the discussion: photodissociation induced by the absorption of single photons. This photoinduced dissociation permits detailed study of molecular dynamics such as breaking of bonds, internal energy transfer, and radiationless transitions. The availability of laser sources with pulses of a few tens of femtoseconds today opens a direct, i.e. real-time, view on this phenomenon.

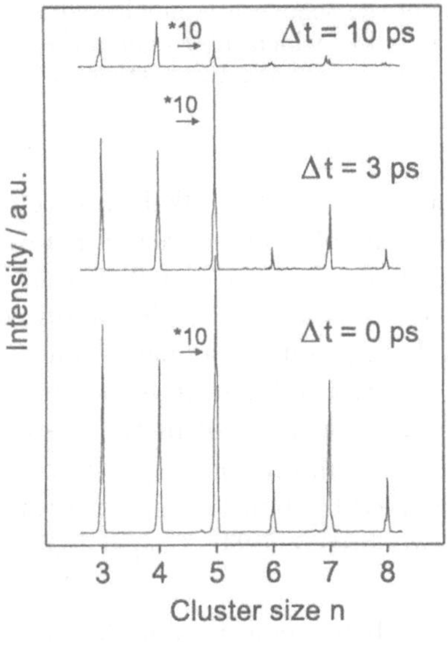

Fig. 4.1. Mass spectra of $Na_{n=3...9}$ for three different delay times Δt obtained, by TPI spectroscopy with sub-50 fs laser pulses. the energy of the pump pulses is 2.93 eV and that of the probe pulses is 1.46 eV

In Fig. 4.1 mass spectra of $Na_{n=3...9}$ excited with laser pulses of sub-50 fs length, measured for three different delay times Δt, are shown. It is clearly visible that for each cluster size the ion intensity is dramatically reduced within a few picoseconds. A precise quantitative determination of the decay times can only be obtained by real-time observation of these photoinduced dissociation processes. The subsequent probe pulse allows the detection of those prepared systems which have not yet been broken.

Figure 4.2 sketches the method of monitoring the real-time dynamics of dissociative systems. Preparing a wave packet on a repulsive (dissociative) PES results in fragmentation of the system, causing a fast depopulation of the excited state.

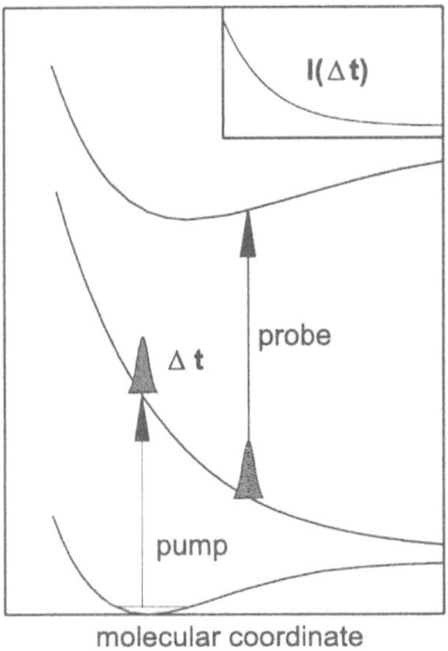

Fig. 4.2. Excitation scheme for probing bound–free transitions. A wave packet is prepared on a repulsive PES. Hence, initially the photoion of the mother aggregate appears within an MPI process. The real-time behavior of the sequence is monitored with the probe pulse, which interrogates the excited system by ionizing the particle after a variable time

Conventional spectroscopy gives clear hints that a few alkali molecules and clusters can act as model systems for ultrafast photodissociation. For example, stationary and nanosecond-spectroscopy on the Na_3 C [369, 374, 393] and D states [124, 393] has already indicated the onset of the photoinduced fragmentation. With increasing size of alkali aggregates fragmentation becomes even more important. As a general result, nondissociative electronic excitation processes have not yet been observed for any free metal clusters larger than the trimers [369]. An alternative to conventional spectroscopy of such 'bound–free' transitions was first provided by depletion spectroscopy (DS)[1] applied to various alkali molecules and clusters [124]. Deeper insight

[1] The technique of depletion spectroscopy can briefly be summarized in the following way. A laser with energy E_1 continously ionizes the ground state population of a molecule or cluster, while a second laser with tunable energy E_2 is used to excite the aggregates from the ground state into a predissociated state of interest. Whenever a predissociated state is resonantly excited by the second laser a depletion of the ground state is directly manifested in a drastic decrease of the

into the real-time dynamics of such photoinduced cluster fragmentation, however, is obtained by ultrafast observation schemes as indicated in Fig. 4.2. The aggregates are electronically excited with ultrashort laser (pump) pulses into the predissociated[2] state. There, they might oscillate a few times (as in the case of K_3 described in Sect. 3.2.5) and then dissociate, or they might dissociate directly as sketched in Fig. 4.2. The real-time behavior of the sequence is monitored with the probe pulse, which interrogates the excited system by ionizing the particle after a variable time.

This type of ultrafast dynamics will now be investigated in detail for several alkali aggregates, beginning with the model molecule Na_3 excited to its C state (Sect. 4.1) and D state (Sect. 4.2). The ultrafast fragmentation of sodium clusters (Sect. 4.3) and potassium clusters (Sect. 4.4) rounds off these studies.

4.1 Predissociation of Na_3 C State

The sodium trimer excited to the electronic C state can be regarded as a fascinating model system, which manifests ultrafast predissociation dynamics. While stationary and nanosecond-pump&probe spectroscopy gave the first hints that this excited state photodissociates rather fast, real-time TPI spectroscopy opens a window to directly observe these ultrafast processes. But let us first start with a short review of the spectroscopy of this excited electronic state.

Spectroscopic Basics of Na_3 C. In the late 1980s special interest was focused on the $C(2)^2E''$ state (in D_{3h} symmetry) of Na_3. Energy-resolved spectroscopy allowed the observation of lower vibrational levels of this electronic state by means of TPI, whereas the upper levels require the use of DS to probe dissociative states [369, 374, 393]. The spectrum of the C state is characterized by a vibrational band structure with pseudorotational features, as shown in Fig. 4.3. These investigations confirmed the C state to be partially predissociated. Therefore, the dissociation channel was proposed to be the main relaxation process for states higher in energy than the C state. This could also be demonstrated for the D state by the depletion technique with a few nanoseconds time resolution[3] [375], as well as for Rydberg states close to the ionization limit [124].

---

ion signal (see e.g. Fig. 3.36). The effect is most efficient when the predissociating transition is saturated.

[2] Ordinary, well-behaved dissociation occurs when a molecule is excited to an electronic state that posesses more energy than the separated fragments. An example of such a transition is shown in Fig. 4.2. Predissociation is dissociation that occurs in a transition before the dissociation limit is attained, hence its name.

[3] In Sect. 4.2 the D state is analyzed with 10 fs time resolution using real-time TPI.

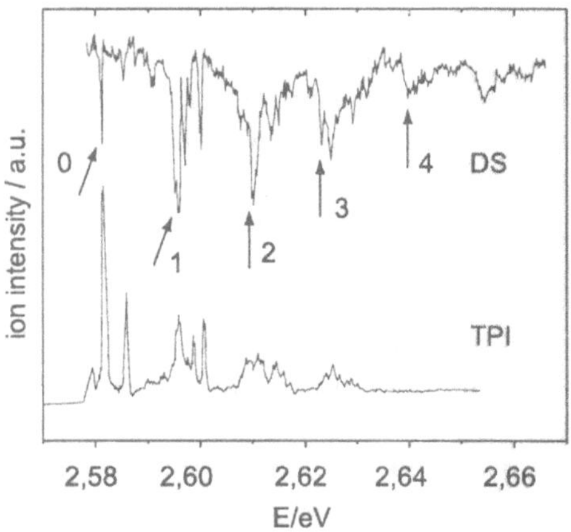

Fig. 4.3. Highly resolved TPI (*lower*) and depletion spectrum (DS, *upper*) of the Na_3 C state (according to [374]). The vibrational bands are labeled $v = 0, 1$ to 4, but note that v is not a vibrational quantum number

The lifetime τ of the C state was measured by nanosecond-time-resolved resonant TPI spectroscopy employing laser pulses of 5 ns duration [374]. The excitation laser was tuned to a given vibrational level of the C state and the ion signal was recorded as a function of the time delay between the exciting and ionizing laser pulses. The jitter of the delay was about 2 ns. The lifetimes of the first two bands were found to be 7 ± 3 ns, quite similar to the radiative lifetimes of the Na_2 A state, 12 ns [394], and of the Na_2 , 7 ns [395]. To calculate the radiative lifetime τ_{rad} of the C state Broyer et al. [374] used a shell model [68, 396]. They estimated τ_{rad} to be 12 ns, slightly larger but compatible with the measured lifetime for these two lower bands. Taking into account the approximate nature of their model and the much more complicated character of the Na_3 eigenfunctions, they remarked that their experiment could not exclude a possible small predissociation for the first two bands. This means that, owing to the branching ratio

$$\frac{1}{\tau} = \frac{1}{\tau_{\text{rad}}} + \frac{1}{\tau_{\text{frag}}}, \tag{4.1}$$

with a fragmentation probability $1/\tau_{\text{frag}}$, the radiative lifetime of the C state could even be slightly greater than the measured lifetime. Later, Bonačić-Koutecký and coworkers estimated the radiative lifetime of the C state to be 6 ns [72] by employing configuration-interaction (CI) studies of excited states. Hence, the main relaxation channel for these lower bands was deduced to be – analogously to the A, B, andB′ states of Na_3 (see Fig. 3.36) – mainly determined by radiative decay.

For all other bands the lifetimes could not be measured by the nanosecond pump&probe technique but were estimated to be less then 5 ns. This suggested that predissociation occurs for at least the highly excited vibrational levels.

Picosecond Real-Time TPI Spectroscopy of Na_3 C. As is shown now, the application of much shorter laser pulses opens a window to ionize the excited Na_3 clusters before most of them can break into Na_2 and Na fragments. Employing the picosecond pump&probe technique allows the measurement of the real-time TPI signal of the vibrational bands of the C state with about 1000 times better time resolution. This enables the study of the fragmentation dynamics in real time and provides a deeper insight into the trimer's suggested onset of predissociation. The mechanism of pump photon ($E_{\text{pump}} = h\nu_{\text{pump}}$) absorption by an observed Na_3 cluster is simply given by (4.2a). Then, either the excited C state (*) can dissociate and two fragment products are found (4.2b), or by absorbing a probe photon ($E_{\text{probe}} = h\nu_{\text{probe}}$) the cluster can be ionized (4.2c).

$$\mathrm{Na}_3 + h\nu_{\text{pump}} \longrightarrow \mathrm{Na}_3^* \tag{4.2a}$$

$$\mathrm{Na}_3^* \longrightarrow \mathrm{Na}_{3-m}^* + \mathrm{Na}_m \quad (m = 1, 2) \tag{4.2b}$$

$$\mathrm{Na}_3^* + h\nu_{\text{probe}} \longrightarrow \mathrm{Na}_3^+ + \mathrm{e}^- . \tag{4.2c}$$

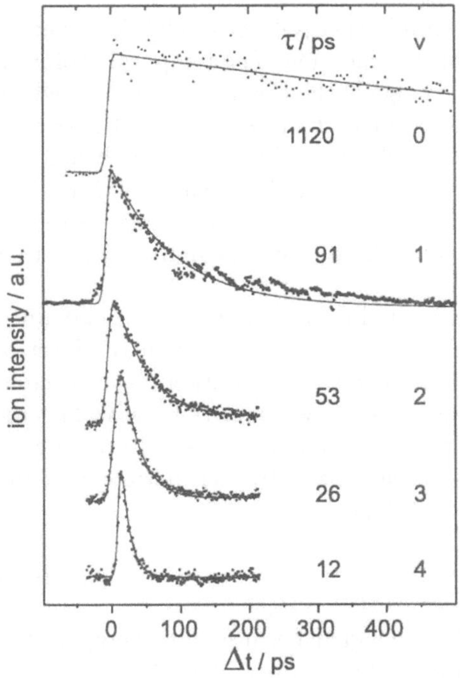

Fig. 4.4. Real-time TPI spectra of different vibrational bands of the Na_3 C state compared with the fit function $f(t) = N_0 \exp(-t/\tau)$ (taken from [392]). The lifetimes τ of the different vibrational bands were obtained by a least-squares fit procedure

The different photon energies for excitation and ionization of the trimer, combined with the request for picosecond temporal resolution, required[4]

[4] Today the author would prefer an arrangement with a titanium sapphire laser pumping synchronously an OPO (see Sect. 2.1.1).

two synchronized titanium sapphire lasers (picosecond-configuration) (see Sect. 2.1.1). This provides independent wavelength tuning of the pump and probe laser trains. To excite the different vibrational levels of the Na_3 C state close to 2.6 eV, the output of one of the titanium sapphire lasers was used to generate the second harmonic. With a BBO crystal 30 mW average power was obtained. The ionization pulse generated by the second laser was tuned to the energy of 1.46 eV and was held fixed for all measurements.

Figure 4.4 displays the measured temporal evolution of the Na_3 ion signal for several excitation energies E corresponding to the observed vibrational bands (see Fig. 3.36). As stated in [393], there is at present no complete nomenclature for these bands. Therefore, to use a simple assignment these vibrational bands are labeled by $v = 0$ to 4, but note that v is not a vibrational quantum number. Each of the recorded real-time ion signals shows a rapidly reached pronounced maximum followed by a fast reduction in intensity. With increasing excitation energy this decay becomes even faster. While at the excitation energy of $E = 2.58$ eV the decay lasts a few nanoseconds, at $E =$ 2.64 eV it is finished within 50 ps. Besides this, the drastic step between $v = 0$ and $v = 1$ has to be emphasized.

For analysis of these results the simple energy-level model described in Sect. 2.1.1 is now applied with a slight modification. The temporal evolution of the ion signal reflects the transient change of the C state population, which can be changed either by radiative decay or by fragmentation of the trimer. The ensemble of larger sodium clusters which are in the cluster beam and which can also interact with the pump pulse might break into Na_3. But the probability of being excited to the C state is nearly zero and, therefore, because of the chosen energy of the probe photons, can be omitted in further discussion.

Thus, the real-time spectra of Na_3 excited to the C state can be explained within the simple energy-level model (see Sect. 2.2.2) shown in the inset of Fig. 4.5. The pump pulse will generate an initial population N_0 of excited trimers within its pulse width. Owing to radiative transitions with lifetime τ_{rad} and fragmentation with lifetime τ_{frag}, this population will decrease with the progress of time t. The decay is characterized by the lifetime τ of the excited ensemble of trimers given by (see 4.1)

$$\tau = \frac{1}{1/\tau_{\mathrm{rad}} + 1\tau_{\mathrm{frag}}}. \tag{4.3}$$

Radiative decay will reduce the real-time signal because the energy of the probe pulse will be not sufficient to ionize these trimers directly. In the case of fragmentation the excited sodium trimer will break into, for example, an excited dimer (Na_2) and a monomer (Na), reducing the generated population as well. Therefore, the decreasing real-time signal can be described by the decay function

$$f(t) = \ell(t) * n(t), \tag{4.4}$$

obtained in Sect. 2.1.1 on the basis of the simple energy-level model. The temporal change of the population is given by

$$n(t) = N_0 \exp\left(-\frac{t}{\tau}\right). \tag{4.5}$$

To take account of the temporal width of the laser pulses the transient spectra are compared with the convolution function, where $\ell(t)$ is the overall system response of our measuring system, which is represented by the cross correlation of the laser pulses.

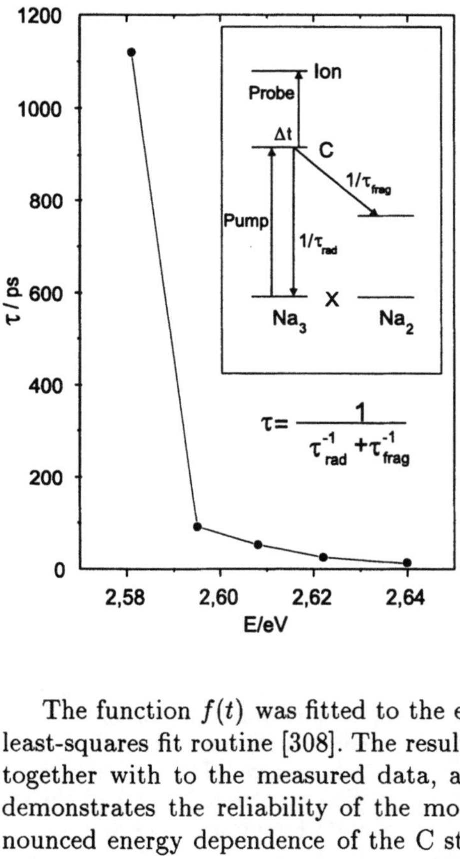

Fig. 4.5. Lifetimes of Na_3 C state's vibrational bands versus excitation energy E. The schematic energy-level system represents the excitation, ionization, fragmentation, and radiative processes of the Na_3 C state. X indicates the ground state. $1/\tau_{frag}$ and $1/\tau_{rad}$ describe the probability of fragmentation and radiative decay respectively of the excited band. Δt marks the time delay between the pump and probe laser pulses (taken from [392])

The function $f(t)$ was fitted to the experimental data $I(t)$ by means of a least-squares fit routine [308]. The resulting fits are shown as lines in Fig. 4.4 together with to the measured data, and are in excellent agreement. This demonstrates the reliability of the model used. Figure 4.5 shows the pronounced energy dependence of the C state's lifetime. The drastic change of τ from $v = 0$ to $v = 1$ is a strong indication that the transition from $v = 0$ to $v = 1$ can be assigned to the threshold of predissociation, i.e. where the excited trimer begins to dissociate with higher probability than that for decay to, for example, the electronic ground state by emission of a photon. The estimated decay times τ are listed in Table 4.1. With growing excitation energy E the lifetime τ of the C state is drastically reduced. Assuming the radiative lifetime – as calculated by [72] – to be 6 ns for all excited bands of the C state, the fragmentation time constant τ_{frag} can be calculated from the measured lifetimes τ. While for $v = 0$ the fragmentation time constant is determined to be $\tau_{\text{frag}} = 1.38$ ns and is comparable with the radiative lifetime for all higher bands $v = 1$ to 4, the fragmentation process clearly dominates over the radiative processes of the C state. Here, the measured lifetimes can be identified with τ_{frag}. The dissociative process is about 100 times faster when exciting the band $v = 4$ compared to the band $v = 0$.

Table 4.1. Lifetimes τ of different vibrational bands v of the Na_3 C state after photoexcitation with energy $E = E_v$

v	0	1	2	3	4
E/eV	2.578	2.590	2.608	2.622	2.640
τ/ps	1120	91	53	26	12

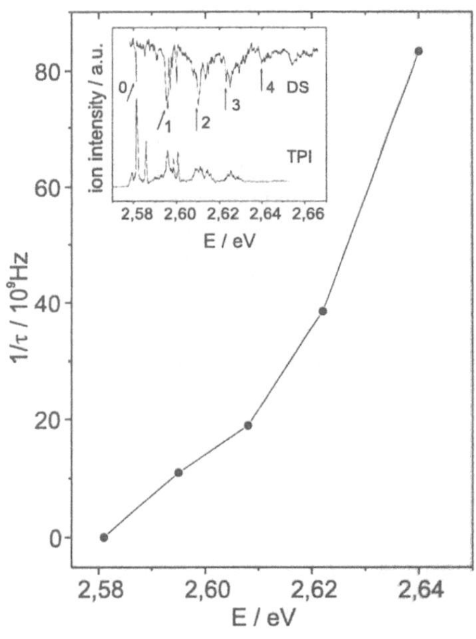

Fig. 4.6. Fragmentation probability of Na_3 C state's vibrational bands versus excitation energy E. Inset: Highly resolved TPI (*lower*) and depletion spectrum (*upper*) of theNa_3 C state (according to [374])

For clusters and their photoinduced dissociation their stability is an important characteristic parameter. Therefore, to introduce a suitable language for the following sections, the stability of the excited trimer is now – even more vividly – described by the fragmentation probability $1/\tau_{\text{frag}}$. With the assumption made above, $1/\tau_{\text{frag}}$ can be identified with the measured value of $1/\tau$ for $v = 1$ to 4, while for $v = 0$ we have $1/\tau_{\text{frag}} = 0.73 \times 10^9$ Hz. Figure 4.6 shows the dependence of the fragmentation probability of the Na_3 C state the on the excitation pump energy. The non-linear increase with growing excitation energy once again documents the sudden change of the trimer's stability in this region. Taking into account the fact that the C state is energetically surrounded by the B state showing no fragmentation (Sects. 3.2.2, 3.2.4 and [68, 379, 380]) and the D state with ultrafast fragmentation within several hundreds of femtoseconds (Sect. 4.2 and [397]), this behavior strongly supports the thesis that the onset of the trimer's predissociation is located close to the band labeled $v = 0$. So, the Na_3 C state seems to be the threshold where the radiative processes occurring on a nanosecond timescale [72] become strongly dominated by the fragmentation processes.

4.2 Excited D State of Na_3 Molecules

While the C state of Na_3 can be partially studied by resonant TPI, the energetically higher-lying D state (418 to 432 nm) can only be observed by depletion spectroscopy, as shown in Fig. 4.7. The spectra, measured by Broyer et al. [124] in 1986, reveal a pronounced, nearly harmonic progression with energy spacings of 81.5 to 88 cm^{-1}. Since no resonant TPI signal is detected, the predissociation lifetime of the D state should be significantly below 1 ns over the whole spectrum.

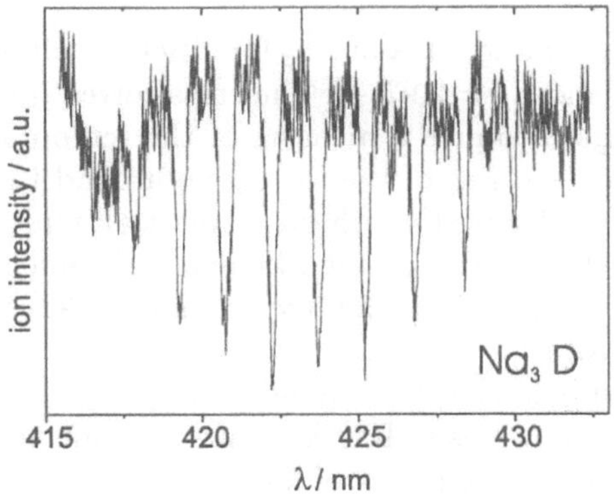

Fig. 4.7. Depletion spectrum of Na_3 D (taken from [124])

In the first pump&probe experiments with picosecond time resolution the real-time spectra of the Na_3 D state (Fig. 4.8) reveal a fast exponential decay caused by ultrafast photo-induced dissociation [306]. This behavior could be well explained with the simple fragmentation model described in Sect. 2.2.2. With femtosecond time resolution it was expected to observe the photodissociation process with more detail in the recorded transients. Especially, the energy dependence of the ultrashort fragmentation process should allow deeper insight into the fragmentation dynamics within a photoexcited molecular beam.

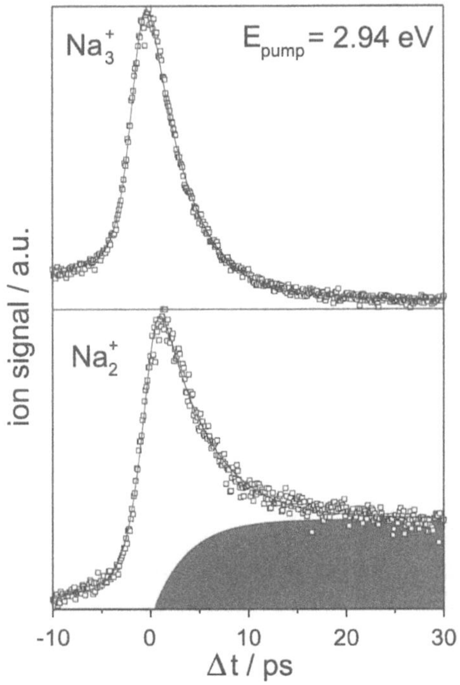

Fig. 4.8. Real-time spectra of Na_3 D excited with laser pulses of 1.4 ps pulse width at $E_{pump} = 2.94$ eV compared with the fragment signal Na_2. The main peak is due to the original Na_2 in the molecular beam. The shaded area represents the amount of Na_2 generated by fragmentation. The lines are least-squares fits obtained within the simple energy level model (taken from [306])

Figure 4.9 shows the measured real-time spectra of Na_3 for different pump energies $E_1 = 2.97$ eV, $E_2 = 2.94$ eV, and $E_3 = 2.92$ eV. The cross correlation of the laser pulses is added to the figure to give a measure of the temporal resolution. For $E_1 = 2.97$ eV, after a steep rise of the ion signal around the zero-of-time, a decrease of the signal is observed which does not at all represent an exponential decay as expected from the simple picosecond fragmentation model [398]. At the excitation energies $E_2 = 2.94$ eV and $E_3 = 2.92$ eV the real-time spectra present two maxima.

To describe this more complicated behavior, the extended fragmentation model given in Sect. 2.2.2 is utilized as a basis. It takes into account the population density and fragmentation characteristics of Type I and Type II clusters.[5] Since for delay times $\Delta t < 0$ and $\Delta t > 0$ the pump and probe pulses play different roles a further fragmentation channel has to be added to the extended model. Hence, the model used here contains four different fragmentation processes with four time constants τ_0, τ_1, τ_2, and τ_3. Here, τ_0 characterizes the fragmentation behavior of the relevant Type I clusters,

[5] As a reminder of the two types of observed clusters described by the extended model, direct fragmentation of the clusters excited by the pump pulse within an average time τ_0 is called Type I fragmentation. Increasing population due to fragmentation of larger clusters into the recorded mass channel with a decay constant τ_1, followed by a further fragmentation of these clusters with time constant τ_2, is called Type II fragmentation.

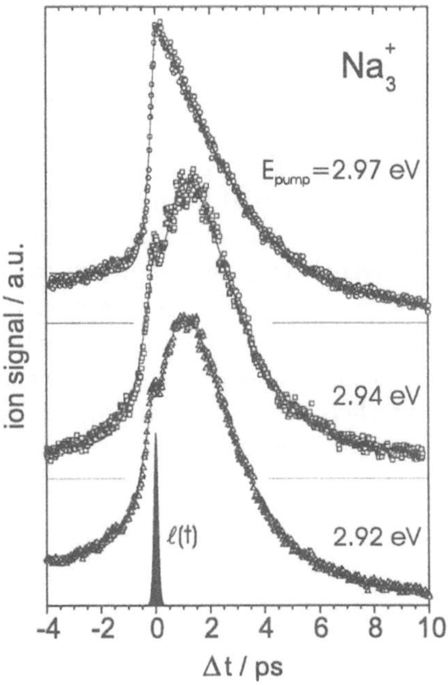

Fig. 4.9. Real-time spectra of Na_3 D excited at three different excitation energies E_{pump} excited with laser pulses of 30 fs (taken from [397]). The shaded curve is the overall system response $\ell(t)$. The lines are least-squares fits calculated within the extended energy level model

whereas τ_1 and τ_2 are due to Type II clusters. τ_3 describes the fragmentation when the pump and probe pulses are exchanged ($\Delta t < 0$ in Fig. 4.10). For this part of the behavior, the simple energy-level model seems to be sufficient.

In the extended model the transient intensity of the detected ions is proportional to the convolution of the population density $n(t)$ of excited clusters (Types I and II) with the overall system response $\ell(t)$ to the laser pulses:

$$I(t) \propto n(t) * \ell(t) = (n_+(t) + n_-(t)) * l(t). \tag{4.6}$$

For $t \geq 0$ ($E_{\text{pump}} = 2.96\,\text{eV}$, $E_{\text{probe}} = 1.48\,\text{eV}$) the temporal dependence of the excited population density is given by $n_+(t)$, described by the sum of three exponential functions [397]:

$$n_+(t) = \overbrace{N_0 \exp\left(-\frac{t}{\tau_0}\right)}^{Type\,I} + \overbrace{M_0 \left[\exp\left(-\frac{t}{\tau_1}\right) - \exp\left(-\frac{t}{\tau_2}\right)\right]}^{Type\,II} \tag{4.7a}$$

$$= n_1(t) + n_2(t) \tag{4.7b}$$

with weighting constants N_0 and M_0. The first term of this sum ($n_1(t)$) represents the temporal evolution of the relevant population density of Type I clusters, whereas the second term ($n_2(t)$) is ascribed to the Type II ions.

For $t < 0$ ($E_{\text{pump}} = 1.48\,\text{eV}$, $E_{\text{probe}} = 2.96\,\text{eV}$) no Type II fragmentation could be observed. Hence, the transient population density for this two-photon process is given by a single exponential decay, namely

$$n_{-}(t) = N_3 \exp\left(-\frac{t}{\tau_3}\right). \tag{4.8}$$

N_3 is a weighting factor, reflecting the initial population density of excited states for $E_{\text{pump}} = 1.48\,\text{eV}$.

To determine the fragmentation probability of Type I clusters, the ion signal of Type II ions has to be separated and subtracted from the observed signal. Therefore, one has first of all to deconvolute the transient ion signal $I(t)$. This is done by convoluting the overall system response with four exponential functions using a least-squares fit algorithm. Then the contribution of the Type II ions has to be subtracted. The last step is to obtain separately the transient population densities $n_1(t)$ and $n_{-}(t)$ of the Type I clusters for $t \geq 0$ and $t < 0$ respectively, by subtracting the other contribution. The result of this procedure is two curves, each decreasing with a single exponential decay. The decay constants are τ_0 and τ_3.

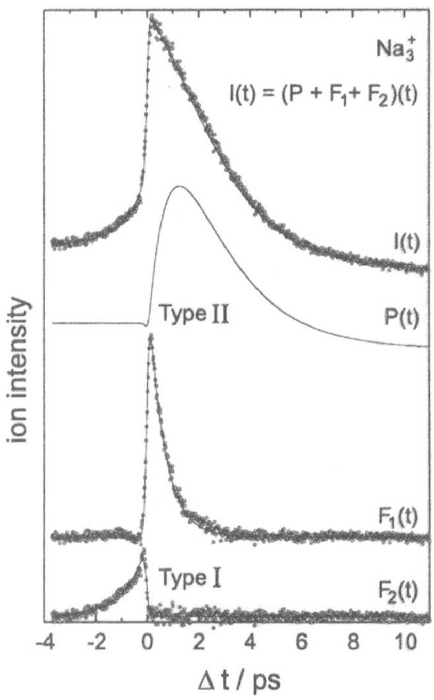

Fig. 4.10. Recorded real-time TPI signal $I(t)$ for Na_3 excited with $E_{\text{pump}} = 2.96\,\text{eV}$ ($\Delta t \geq 0$) and $E_{\text{pump}} = 1.48\,\text{eV}$ ($\Delta t < 0$) (taken from [263]). Owing to the different excitation, ionization, and fragmentation processes of the cluster ensemble in the beam, $I(t)$ is the sum of three processes: $I(t) = P(t) + F_1(t) + F_2(t)$. $F_1(t)$ and $F_2(t)$ describe the photofragmentation of the Type I clusters, whereas $P(t)$ represents Type II fragmentation. Dots, experimental data; lines, fitted function due to the fragmentation model explained in the text (Δt, delay time)

Figure 4.10 presents, as an overview, the separated contributions of $I(t)$ obtained by this algorithm. For $t \geq 0$ (Type II) the contribution is

$$P(t) = l(t) * n_2(t), \tag{4.9}$$

for $t \geq 0$ (Type I)

$$F_1(t) = l(t) * n_1(t) \propto l(t) * \exp\left(-\frac{t}{\tau_0}\right), \tag{4.10}$$

and for $t < 0$ (Type I)

$$F_2(t) = l(t) * n_-(t) \propto l(t) * \exp\left(-\frac{t}{\tau_3}\right). \tag{4.11}$$

Hence, $I(t)$ is given by the sum

$$I(t) = P(t) + F_1(t) + F_2(t). \tag{4.12}$$

The comparison of least-squares fit functions and experimental data (Fig. 4.9) shows excellent agreement. It has to be pointed out that although the fit routine uses four exponential functions, for negative delay times only one of these is essential. The remaining three describe the behavior for positive delay times only. Here, again only one exponential function is necessary in the case of clusters with an even number of atoms. The decay constants τ_0, τ_1, τ_2, and τ_3 obtained by a least-squares-fit procedure are listed in Table 4.2. One can easily compare the values of the Type I fragmentation with the vibrational data obtained from depletion spectroscopy. With the energetic spacing of the vibrational levels in the D state ($\omega_0 = 85\,\mathrm{cm}^{-1}$) the period of a wave packet amounts to 390 fs. Hence, the fragmentation process takes place within about a single vibration of the excited trimer.

Table 4.2. Decay constants τ_0, τ_1, τ_2, and τ_3 as a function of the pump energy E obtained by a least-squares-fit procedure

E/ eV	2.92	2.94	2.97
τ_0/ fs	480	260	640
τ_1/ fs	770	870	870
τ_2/ fs	2370	1680	2000
τ_3/ fs	870	440	280

The fragmentation times of the sodium D state and the fragmentation of larger clusters into Na_3 fragments show a strong dependence on the energy of the exciting pump pulses. It is obvious that explaining this dependence is out of the range of the present model. The energy dependence may be correlated to various channels of vibronic or electronic predissociation. For a deeper interpretation it is necessary to know the exact potential-energy surfaces and intersystem crossings for the interacting particles and to take into account wave packet motions on these PESs.

4.3 Small Sodium Clusters

Sodium clusters seem to be the most extensively studied species in cluster science. From a theoretical point of view they can be regarded as a prototype system with metalic bonding, revealing a pronounced delocalization of charge even for the smaller sizes. In the case of very small sodium clusters ab initio configuration-interaction (CI) calculations [399, 400] are preferred, whereas for larger clusters the jellium model leads to good agreement with experimental results [401]. Fundamental experimental results have been obtained by laser [127, 107, 402] and mass spectroscopy [403, 404].

Here, studies of the ultrafast photodissociation dynamics of excited electronicstates of Na_4 up to Na_{10} by means of femtosecond real-time pump&probe spectroscopy are described. The two-color TPI technique with laser energies of 1.48 eV and 2.96 eV (second harmonic generation from the output pulses of the femtosecond configuration of the titanium:sapphire laser) was employed. In Fig. 4.11 a the real-time evolution of the ion signal is presented for different cluster sizes. The intensity of the ion signal at $E_{\text{pump}} = 2.96\,\text{eV}$ (positive delay times) is, for all sizes of the Na_n clusters, higher than at $E_{\text{pump}} = 1.48\,\text{eV}$ (negative delay times). This corresponds to the higher density of electronic states in the region around 3 eV. The decay of the ion yield in each curve might have two reasons: first, the ultrafast fragmentation of the excited cluster, and second, a reduction due to a rapid intramolecular vibrational redistribution of the excited system. However, since no wave packet propagation indicated by oscillations superimposed on the decay curves (as seen e.g. for the dissociative K_3 molecule excited to its B state, discussed in Sect. 3.2.5 and [260, 381]) are observed, it is concluded, that fragmentation is the only reason for the ultrafast decay.

As a first approximation Kühling et al. [405] supposed that excited sodium clusters break into a monomer or dimer and the respective daughter cluster. To split off a monomer or a dimer, an energy of 0.6 eV or 0.9 eV respectively is necessary [406]. Hence, taking into account the energy balance of the Type II processes, the ionization of the fragments needs – in nearly all of the cases examined – two photons of the probe pulse. Starting with the excited cluster, the detailed Type II processes studied here are listed below. The processes calculated for $E_{\text{pump}} = 2.96\,\text{eV}$ are based on the data given in [90, 406]. The results are summarized by the following reaction equations

$$Na_3^* \longrightarrow Na_2^* + Na \xrightarrow{2h\nu} Na_2^+ \tag{4.13a}$$

$$Na_4^* \longrightarrow Na_3^* + Na \xrightarrow{h\nu} Na_3^+ \tag{4.13b}$$

$$Na_5^* \longrightarrow Na_4^* + Na \xrightarrow{2h\nu} Na_4^{+*} \longrightarrow Na_3^+ + Na \tag{4.13c}$$

$$Na_6^* \longrightarrow Na_4^* + Na_2 \xrightarrow{2h\nu} Na_4^{+*} \longrightarrow Na_3^+ + Na \tag{4.13d}$$

$$Na_7^* \longrightarrow Na_6^* + Na \xrightarrow{2h\nu} Na_6^{+*} \longrightarrow Na_5^+ + Na \tag{4.13e}$$

$$Na_8^* \longrightarrow Na_6^* + Na_2 \xrightarrow{2h\nu} Na_6^{+*} \longrightarrow Na_5^+ + Na \tag{4.13f}$$

$$Na_9^* \longrightarrow Na_8^* + Na \xrightarrow{2h\nu} Na_8^{+*} \longrightarrow Na_7^+ + Na. \tag{4.13g}$$

These equations nicely demonstrate that Type II processes will be detected in the detection channel of Na_3^+, Na_5^+, and Na_7^+. This result is as well

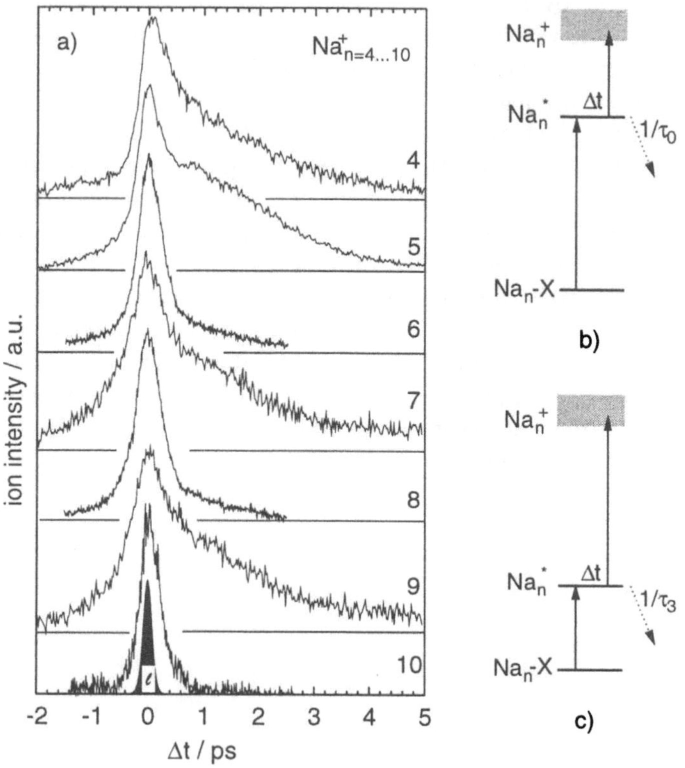

Fig. 4.11. Real-time evolution of the ion signals for $Na_{n=4...10}$ (taken from [263])(**a**) and energy level schemes to describe the temporal evolution of the Na_n^+-signal (X = ground state) for $\Delta t > 0$ $E_{pump} = 2.96\,eV$ and $E_{probe} = 1.48\,eV$ (**b**) and for $\Delta t < 0$ $E_{pump} = 1.48\,eV$ and $E_{pump} = 2.96\,eV$ (**c**). $1/\tau_0$ and $1/\tau_3$ indicate the fragmentation probabilities of an excited state Na_n^* for different excitation (E_{pump}) and ionization energies (E_{probe})

in good agreement with a photofragmentation analysis by Hertel, Schulz, and coworkers [107]. They found, in an experiment with size-selected neutral sodium clusters, that during photoionization mainly ionic fragments with an odd number of atoms and an even number of electrons were formed. Therefore, the real-time ion signal of the odd-numbered clusters in the present study should deviate to some extend from a single exponential decay, whereas for even-numbered clusters a single exponential decay is expected.

The real-time spectra indeed reveal this predicted behavior (see Fig. 4.11). The spectra were analyzed within the extended fragmentation model (see Sect. 2.2.2). For clusters with an odd number of atoms a fairly large number of Type II clusters is observable(Fig. 4.12 a), which is in excellent agreement with the reaction equations given above. The real-time spectra of clusters with an even number of sodium atoms show a decay which can be described by the convolution of the overall system response with a single exponential function (Fig. 4.12 b). The relevant fragmentation time constants τ_0 and τ_3 were obtained by the mathematical procedure described in Sect. 4.2 for the different cluster sizes and are listed in Table 4.3. These values mirror the lifetimes of the photoexcited clusters for two different excitation energies E_{pump}. In Fig. 4.13 the temporal evolution of the pure *Type I* fragmentation (i.e. without Type II processes) is shown in comparison to the least-squares fits for both excitation energies. To obtain these isolated Type I photodissociation dynamics the mathematical filter procredure described in Sect. 4.2 was applied. Figure 4.14 summarizes these results by presenting the fragmentation probabilities $1/\tau_0$ and $1/\tau_3$ as a function of the number n of atoms of the observed cluster. For $E_{\text{pump}} = 1.48\,\text{eV}$ an even–odd alternation (except for $n = 4$) of the photodissociation probability is clearly visible. Even-numbered clusters tend to dissociate faster, whereas the odd-numbered clusters are the more stable candidates with respect to this excitation energy.

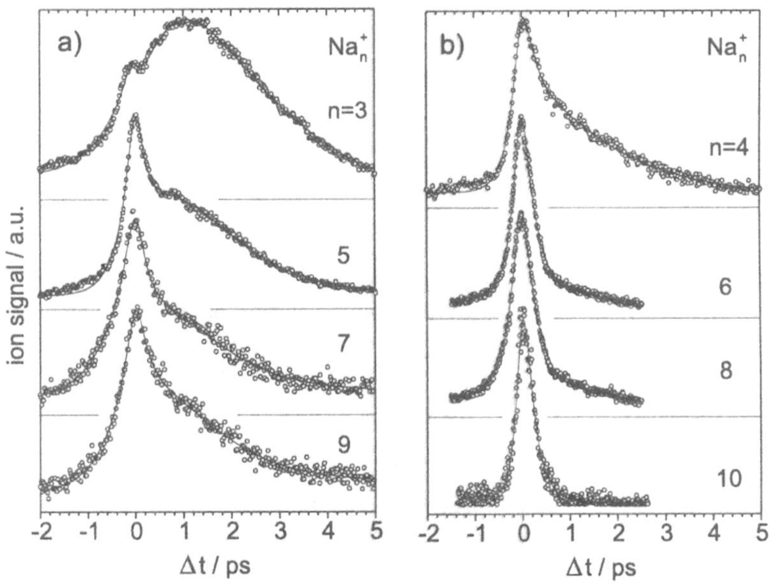

Fig. 4.12. Comparison of real-time spectra (*circles*) and least-squares fits (*lines*) obtained within the extended energy-level model for Na_n clusters built up by **(a)** an odd and **(b)** an even number of sodium atoms (taken from [405]). For $\Delta t > 0$ $E_{\text{pump}} = 2.96\,\text{eV}$ and $E_{\text{probe}} = 1.48\,\text{eV}$, and vice versa for $\Delta t > 0$

Besides the distinct even–odd alternation for both excitation energies, an obvious change with $n = 5$ appears. For $E_{\text{pump}} = 1.48\,\text{eV}$ the even–odd alternation is drastically enhanced. Here, compared to Na_3, the fragmentation probability of Na_{10} is about 8 times bigger. For $E_{\text{pump}} = 2.96\,\text{eV}$ the frag-

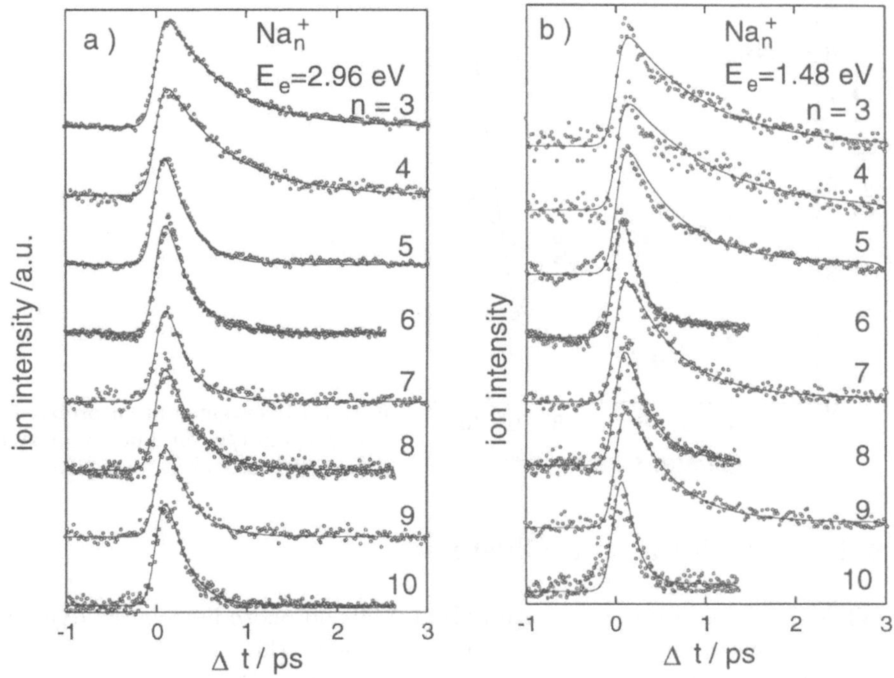

Fig. 4.13. Temporal evolution of the pure Type I photofragmentation (taken from [263]). **(a)** $F_1(t)$ for $\mathrm{Na}^+_{n=3\ldots10}$ excited at $E_{\mathrm{pump}} = 2.96\,\mathrm{eV}$ and **(b)** $F_2(t)$ for $\mathrm{Na}^+_{n=3\ldots10}$ excited at $E_{\mathrm{pump}} = 1.48\,\mathrm{eV}$. *Dots*, experimental data; *lines*, fitted single exponential function

Table 4.3. Lifetimes τ_0 and τ_3 of Na_n clusters after photoexcitation with energy $E_{\mathrm{pump}} = 2.96\,\mathrm{eV}$ and $E_{\mathrm{pump}} = 1.48\,\mathrm{eV}$ respectively. The error in determination of the lifetimes is $\Delta\tau = \pm 10\,\mathrm{fs}$

n	3	4	5	6	7	8	9	10
τ_0/fs	600	690	260	270	240	310	280	220
τ_3/fs	970	990	580	170	550	270	510	120

mentation probability shows at $n = 5$ a sudden increase to about 2 times the value at $n = 3$. Hence, for this excitation energy the stability of clusters larger than Na_4 is drastically smaller than that of the smaller ones.

It has to be stated that this simple fragmentation model cannot explain any energy dependence of the recorded data. For other excitation conditions, e.g. as used by Gerber and coworkers in their experiments on Na_n, the model might lose its validity. For example, Gerber and coworkers [71, 132] excited Na_8 close to four surface plasmon resonances at 2.39 eV (518 nm). Therefore, several ultrashort decay processes are involved simultaneously, instead of one in the case discussed here.

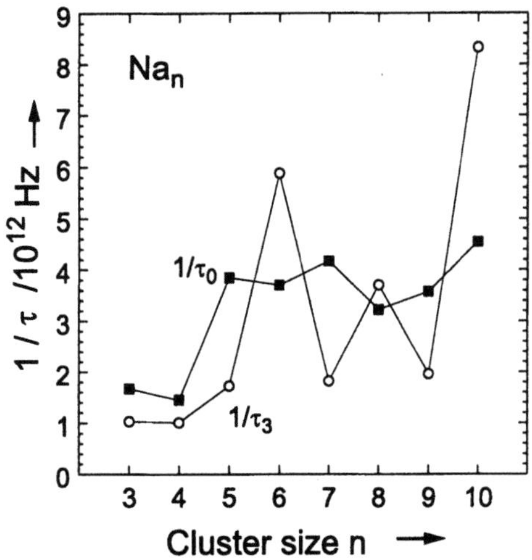

Fig. 4.14. Fragmentation probabilities $1/\tau_0$ and $1/\tau_3$ for different cluster sizes n of Na_n (taken from [263])

4.4 Small Potassium Clusters

While the dipole absorption features [114, 124] and photodissociation dynamics of small sodium clusters are rather well known [112, 113, 126, 127, 407–409], there is very little knowledge about potassium clusters larger than the dimer. The lack of experimental data might be caused by ultrafast fragmentation processes within the potassium clusters, so that conventional stationary spectroscopic techniques might fail. Hence, the goal of this section is to determine the photodissociation probability of small potassium clusters as a function of cluster size as well as excitation energy.

At present, only preliminary unpublished results on the internal electronic structure of K_4 and K_8 exist [410]. From these ab initio calculations employing the configuration-interaction method, one finds that K_4, in the case of rhombic geometry (with D_{2h} symmetry), has a rather strong oscillator strength at about 1.5 eV (with irreducible representation 1^1B_{1u}) and at 2 eV (2^1B_{2u}). For 3 eV a bundle of states is observable, each with a rather low oscillator strength. For K_8, with a tetrahedral structure (T_d symmetry), there are two states (1^1T_2 and 2^1T_2) in the region of 1.5 eV. For the calculated structure, an oscillator strength 3 to 4 times higher could be estimated for an electronic state (3^1T_2) at about 2 eV, while for 3 eV a quite similar situation to that for K_4 is found. To ionize the excited potassium clusters one has to reach with the probe pulse at least the ionization threshold, which is between 4 and 4.5 eV for $K_{n=3...9}$ [111].

Special Features of the Experimental Setup. Bréchnignac's investigations [111] encouraged the study of the photodissociation dynamics of K_n for three different excitation energies. Employing a two-color TPI experiment, the potassium clusters were excited at 1.47 eV and 2.94 eV, while with one-color TPI spectroscopy these clusters were excited at 2 eV. Therefore, two different optical setups were utilized, which are called, in the following, 'one-color' and 'two-color' experiments, respectively. In both cases the source for ultrashort pulses was a mode-locked Ti:sapphire laser providing pulses with a duration of 70 fs, a spectral width of 190 cm^{-1}, and an average power of 1.2 W at a repetiton rate of 80 MHz. The energy per pulse was 15 nJ rather low compared to amplified laser systems with a repetition rate of a few Hz or kHz but similar output power.

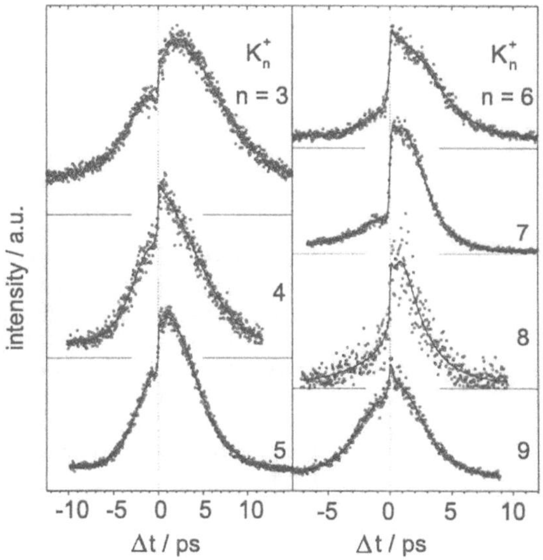

Fig. 4.15. Real-time spectra of the two-color experiment for different cluster sizes n (taken from [307]). For $\Delta t < 0$: $E_{\mathrm{pump}} = 1.47\,\mathrm{eV}$, $E_{\mathrm{probe}} = 2.94\,\mathrm{eV}$; for $\Delta t > 0$: $E_{\mathrm{pump}} = 2.94\,\mathrm{eV}$, $E_{\mathrm{probe}} = 1.47\,\mathrm{eV}$. *Dots*, experimental data; *lines*, fitted curves based on the extended fragmentation model

In the two-color experiments the pulses of the Ti:sapphire laser (1.47 eV) were frequency-doubled by a 1 mm BBO crystal with a conversion efficiency of 15%. A dichroic mirror separated the fundamental from the second harmonic (2.94 eV). For convenience, a positive delay here means the pump beam is the second harmonic at 2.94 eV and the probe beam is the fundamental at 1.47 eV. For negative delay times the energies of the pump and probe pulses are interchanged.

The one-color experiment was performed with pump and probe pulses of the same photon energy (2.00 eV). In this case a synchronously pumped femtosecond optical parametric oscillator was used (see Sect. 2.1.1). At $\lambda =$ 1.3 μm the signal wave's maximum output reached more than 400 mW, corresponding to 20% conversion efficiency. The signal wave was frequency-doubled by a BBO crystal (60 mW). By measuring an interferometric autocorrelation

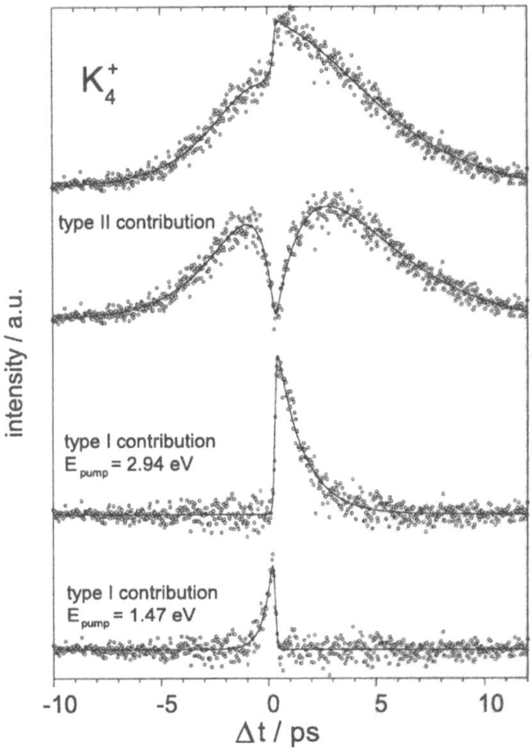

Fig. 4.16. Contribution of the various types of fragmentation to the ion signal in the case of K_4 (taken from [307]). *Dots*, experimental data extracted by the mathematical filter; *lines*, fitted curves extracted by the mathematical filter

with the cluster beam acting as the nonlinear medium, the pulse duration in the interaction zone could be determined to be about 60 fs.

Ultrafast Fragmentation Probability of K_n Clusters. The real-time evolution of the ion signal for the two-color experiment is shown in Fig. 4.15 for different cluster sizes. For all cluster sizes the intensity of the ion signal at $E_{\text{pump}} = 2.94\,\text{eV}$ (positive delay) is higher than at $E_{\text{pump}} = 1.47\,\text{eV}$ (negative delay), possibly corresponding to the higher density of electronic levels in the region of 3 eV.

To describe the shape of the curves, the extended fragmentation model (Sect. 2.2.2) was again applied. The analysis was performed analogously to that for the sodium clusters (Sect. 4.3). The function $n_{\text{ion}}(t)$ was fitted to the experimental data by means of a least-squares routine. This function is shown in Figs. 4.15 and 4.16 as solid lines. Obviously there is an excellent agreement between measured and fitted curves.

The mathematical filter implemented allows the extraction of the relevant Type I contribution from the measured transient spectrum. This part contains the essential information about the laser-excited state. Figure 4.16 shows the different contributions of the ion signal in the case of K_4 obtained by this procedure.

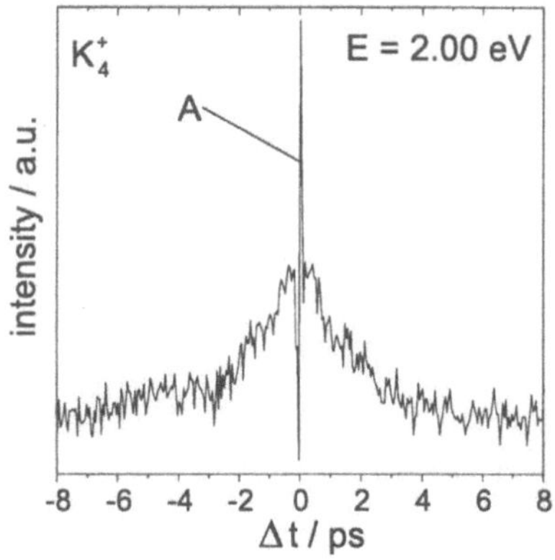

Fig. 4.17. K_4 real-time spectrum from the one-color experiment (taken from [307]). A: fast oscillation at the zero of time point due to the interference of pump and probe pulses

The real-time spectra of the one-color experiments exhibit an interferometric peak overlapping the symmetric shape of the fragmentation curve at $\Delta t = 0$. As an example a measured K_4 spectrum is shown in Fig. 4.17. Necessarily, this narrow peak has to be cut out of the spectra before evaluating the decay times using a least-squares fit algorithm. In order to avoid artefacts the fit was started at the edge of this narrow region, a few femtoseconds wide around the zero of time (Fig. 4.18). The spectra of K_5, K_6, K_8, and K_9 show almost a single exponential decay. Hence, these clusters are strongly dominated by Type I fragmentation and consequently their spectra require no further numerical extraction. The pump&probe spectra of K_3, K_4, and K_7 exhibit a Type II contribution, which, however, is less than that for the corresponding two-color spectra. The evaluation of these spectra was done as described above for the two-color spectra.

Table 4.4. Type I dissociation times τ_1 for different pump energies E and cluster sizes n. The errors are determined as the usual standarddeviation taken over the different measurements. If there is no error indicated, only one transient spectrum has been measured

Cluster size n	τ_1 / ps $E = 1.47$ eV	2.00 eV	2.94 eV
3	0.58±0.03	9.5±1.0	1.09±0.02
4	0.56±0.05	1.83±0.17	1.21±0.11
5	0.50±0.04	1.63±0.17	0.67±0.03
6	0.25±0.02	1.16±0.26	0.51±0.05
7	0.24±0.02	1.22±0.16	0.39±0.02
8	0.75	0.74±0.10	0.24
9	0.39±0.03	1.36±0.15	0.44±0.02

Although three decay times are obtained from each measurement, the main point of interest is the Type I quantity τ_1, because it characterizes

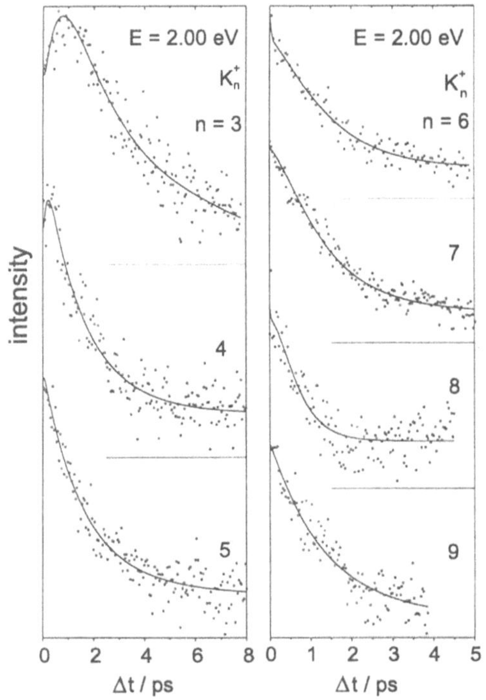

Fig. 4.18. Real-time spectra of the one-color-experiment (taken from [307]). It is $E_{\text{pump}} = E_{\text{probe}} =$ 2.00 eV. *Dots*, experimental data; *lines*, fitted curves based on the extended fragmentation model

the dynamics of the electronic states which are directly populated by the pump pulses. The fragmentation times τ_1 are listed in Table 4.4. These data allow the direct determination of the photodissociation probabilities $1/\tau_1$ for the observed clusters excited at the three different energies. This quantity mirrors the stability of the clusters (Fig. 4.19). For all measured cluster sizes the fragmentation probabilities at $E = 2.00$ eV are smaller than those for the other photon energies. For $E = 2.00$ eV and $E = 2.94$ eV the dependence of photodissociation probability on the cluster size exhibits a similar qualitative course (Fig. 4.19 b). The maximum of $1/\tau_1$ for K_8 is in contrast to an expected higher stability at K_8 predicted by a simple jellium model. At $E = 1.47$ eV, however, the size dependence is totally different. Here indeed we find a minimal photodissociation probability at K_8.

Comparing these results with those presented for $Na_{n=3...10}$ in Sect. 4.3 and [263, 405], several statements can be made. As for the sodium clusters the ratio M_1/N_1 again, although more weakly, shows an even–odd alternation with change of cluster size [411]. Hence, the dissociation mechanism should be rather similar for K_n and Na_n for the measured cluster sizes under similar excitation conditions. However, the overall change of the decay times with the cluster size indicates a major difference. For K_n excited with 2.94 eV there is a quasicontinuous increase of the fragmentation probability $1/\tau_1$ observable up to $n = 8$. In contrast to this, for Na_n $1/\tau_1$ grows drastically while going

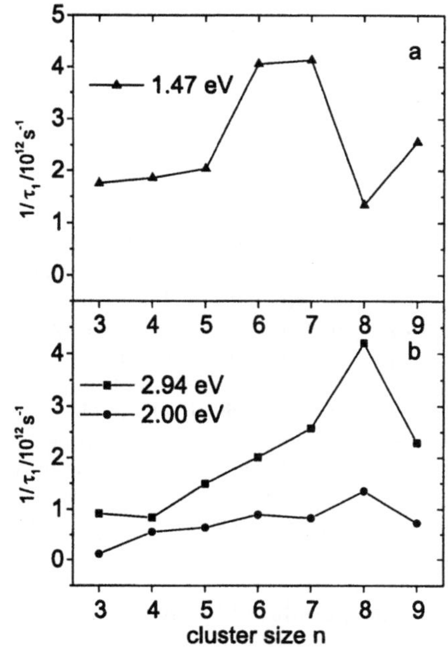

Fig. 4.19. Type I fragmentation probabilities $1/\tau_1$. (a) $E_{\text{pump}} = 1.47$ eV, (b) $E_{\text{pump}} = 2.00$ eV and $E_{\text{pump}} = 2.94$ eV (taken from [307])

from $n = 4$ to $n = 5$ followed by a more or less constant plateau with $1/\tau_1 \approx 3.8 \times 10^{12}$ Hz. Since $1/\tau_1$ characterizes the stability of the excited cluster this indicates a different cohesion of the alkali atoms in K_n and Na_n clusters of equal size n, although they are of similar electronic structure. Perhaps the different sizes of the K and Na atoms result in different preferred geometric structures of the excited clusters, which might be the reason for a different bonding strength. This is well supported by the data for Na_n excited at 1.47 eV (see Sect. 4.3). There a pronounced even–odd alternation of the fragmentation probability, with values between 1.5 and 8×10^{12} Hz being found, which is significantly different to the behavior of K_n as shown in Fig. 4.19a.

A question which is still open is the influence of ionic fragmentation on the measurements described here. This is a rather crucial point since the total photon energy in our experiment ranges from 4 to 4.5 eV and exceeds the ionization potential of the K_n clusters by 0.5 to 1 eV. This is in the range of their fragmentation energies as measured by Bréchignac et al. [111]. Hence, this ionic fragmentation might – as discussed below and illustrated in Fig. 4.20 – contribute to the total ion signal and therefore influence the temporal evolution of the detected ion signal.

A dissociation process, which takes place after the ionization by the probe pulse, cannot be observed directly with the experimental setup used here. Nevertheless there might appear ionic fragments of larger clusters in the recorded mass channel. Figure 4.20 shows as an example, in a scaled energy representation, the pump&probe experiment for excitation at $E_{\text{pump}} =$

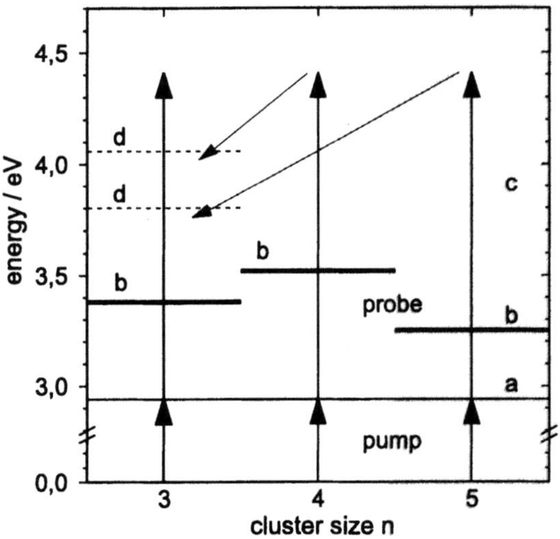

Fig. 4.20. Possible contributions to the ion signal of K_3 caused by ionic fragmentation (taken from [307]). Energy representation of the two-color-experiment. a, pump pulse excited state; b, ionization potential; c, energetic threshold for ionic fragmentation; d, maximum energy of the ionic fragments

2.94 eV. The electronic states of $K_{n=3,4,5}$ (a) are populated by the pump pulse (solid arrow). The photon energy of the probe pulse (solid arrow) exceeds the ionization potential (b), and in the case of K_4^+ and K_5^+, even the limit for ionic dissociation (label c), which has been calculated by Spiegelmann and Pavolini [412]. This means that from an energetic point of view ionic dissociation can occur. The most probable dissociation channels of K_4^+ and K_5^+ (dashed arrows) lead into the K_3^+ signal, as has been measured by Bréchignac et al. [111]. In addition to the (stable) trimer ions created by the probe pulse there might be contributions due to the ionic fragments resulting from K_4^+ and K_5^+. To examine these effects, further measurements with a probe pulse independent of the pump pulse energy are required. For this, the OPO technique opens a gateway and, hence, these experiments will surely be done in the near future.

5. Ultrafast Structural Relaxation

A desideratum of experimental cluster science has been a means to study structures, spectra, dynamics, and reactions of neutral clusters of a single, known mass. One method for doing this with relatively small clusters is deflection by collision with He atoms [413]. Here a new approach is described, which has been used to study silver clusters, Ag_n ($n = 3, 5, 7, 9$). As sketched in Fig. 5.1, the method begins with a beam of mass-filtered, negatively-charged clusters, which are subjected to photodetachment and, after a variable but selected temporal delay, photoionized. The positive ions are then mass-analyzed and collected. The intensity of the positive-ion signal as a function of the delay interval between the two, ultrashort pulses is a measure of the Franck–Condon factor for photoionization of a neutral prepared by a vertical detachment process from a low-lying vibrational state of the negative ion. In this way, the vibrational motion in the neutral, i.e. its ultrafast structural relaxation initiated by the detachment laser pulse, can be directly probed. As defined in Chap. 1, this charge reversal process is referred to as 'NeNePo', **Ne**gative-to-**Ne**utral-to-**Po**sitive.

It is shown here that such photodetachment experiments are a powerful tool to observe and charaterize properties of the transition state of molecules and clusters. Neumark and coworkers nicely demonstrated this in cw experiments [414–417]. The approach using ultrashort laser pulse excitation enables the preparation of the neutral in a superposition of eigenstates. Hence, wave packet dynamics can take place. The corresponding wave packet propagation is probed by the ionization process, which results in the production of the cation and a further photoelectron. Here it is shown that the real-time detection of the cation signal yields new information about the ultrafast nuclear dynamics of the neutral molecules or clusters.

The first studies have been done with molecules and clusters of silver atoms [418], because there is considerable information about them from experiment and from theory. Only the clusters Ag_3, Ag_5, Ag_7, and Ag_9 have been looked at, because their ionization potentials are low enough to ionize them by two-photon absorption with the laser sources available (see Sect. 2.1.2). However, these are in some respects an ideal series because they provide several different structural relationships among the negative, neutral, and positive species [84, 419]. As is shown in Sect. 5.1, the silver trimer can

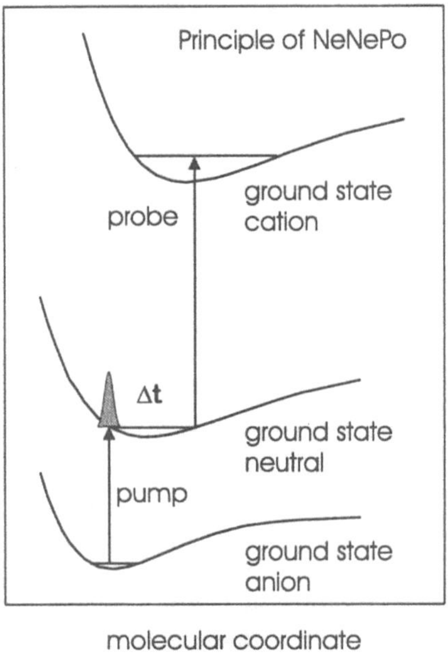

Fig. 5.1. Principle of the NeNePo process. Beginning at the anion's potential-energy surface, an ultrashort pump pulse detaches an electron and prepares a wave packet in the neutral. After a certain delay time Δt a probe pulse photoionizes the neutral. Here, as used in the experiments on Ag_n described here, two probe photons are necessary for ionization of the prepared neutral molecule. The time-dependent signal of the cation's intensity is detected

be regarded as a model system for this new approach to study the ultrafast structural relaxation of this charge reversal process. The first investigations on some larger silver aggregates are described in Sect. 5.2.

5.1 Charge Reversal Process in Ag_3 Molecules

In this section the ultrafast structural redistribution of Ag_3, initiated by a 100 fs laser pulse, is presented[1]. The experiment makes use of a high-intensity cluster anion source, an ion trap, a mass-analyzing detector for cluster cations, and a laser system which produces pairs of ultrashort laser pulses with an adjustable time delay between the pump and the probe laser pulse (for details see Sect. 2.1.2).

Negatively charged, mass-selected silver trimers are caught in a linear quadrupole trap. There, their excess electrons are photodetached with a pulse of radiation in the range of 400 nm, with a duration of nearly 100 fs. The neutrals prepared in this way would remain in the trap for nanoseconds or even milliseconds. However, a second ultrashort pulse of the same radiation, intense enough to induce TPI of the clusters, is directed into the trap after a preselected delay, allowing the neutral to carry out internal motion, from a small fraction of a vibration to many vibrations. The positive ions generated in this way, are then mass-analyzed and collected. Thus, the yield of

[1] A recent extension of the experimental study of Ag_3^- by the 'NeNePo' technique, utilizing two-color excitation, is given by Boo et al. [420].

mass-selected cluster cations produced from mass-selected cluster anions is observed as a function of the delay time between the pump and the probe pulse at fixed wavelength and fluence of these pulses. As is shown, the transient ion signal reveals the internal motions of the atoms of the neutral, monodisperse clusters, particularly in cases in which detachment or ionization or both involve a significant change of geometry.

The first experiments were carried out using the silver trimer anion. Mass-selected Ag_3^- ions were produced with an intensity of about 2 nA and stored in the ion trap. The detachment was performed at wavelengths of 420 nm, 415 nm, 400 nm, and 390 nm, so that one-photon detachment of the anions was possible. The ionization was done nonresonantly using two photons of the same wavelength. The energy of two photons of 420 nm is only slightly above the ionization potential of the silver trimer, thus allowing a very soft ionization.

Using a wavelength of 415 nm, the positive ions were mainly detected when there was a nonzero time delay between the pump and probe laser pulses, confirming that sequential processes of detachment and ionization are involved in the creation of the cations. Remarkably, more than 90% of the cluster cations were detected as trimers, showing that with ultrashort laser pulses, nonresonant multiphoton ionization with very little fragmentation is indeed possible. Nevertheless, small fragment peaks are detectable.

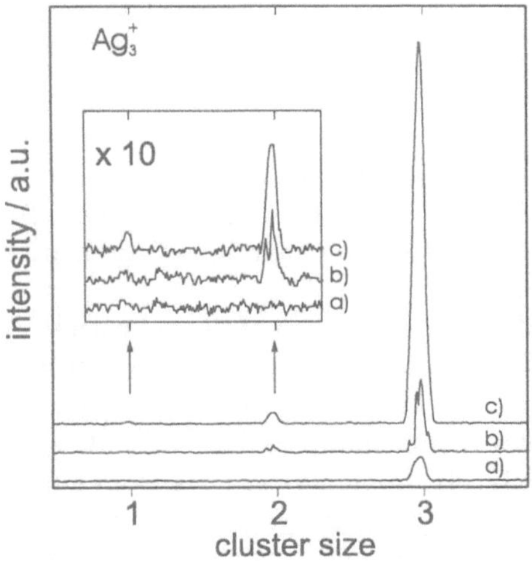

Fig. 5.2. Mass spectra of cationic silver clusters produced from mass-selected Ag anions using **(a)** a single laser pulse, **(b)** two laser pulses with $\Delta t = 0$, and **(c)** two subsequent laser pulses with $\Delta t = 650$ fs ($\lambda = 415$ nm). The structure in **(b)** is due to interference between the pump and probe pulses, and demonstrates that the timing assignment $\Delta t = 0$ is correct (taken from [418])

Both results are clearly visible in Fig. 5.2, which show mass spectra of cluster cations produced using a single laser pulse (Fig. 5.2 a), or using pump and probe lasers at zero (Fig. 5.2 b) or finite (Fig. 5.2 c) time delay. It is clearly apparent that neither a single laser beam nor pump and probe laser pulses with zero time delay produce silver cluster cations efficiently. Rather, the ion signal reaches a maximum when the time delay Δt is approximately 650 fs

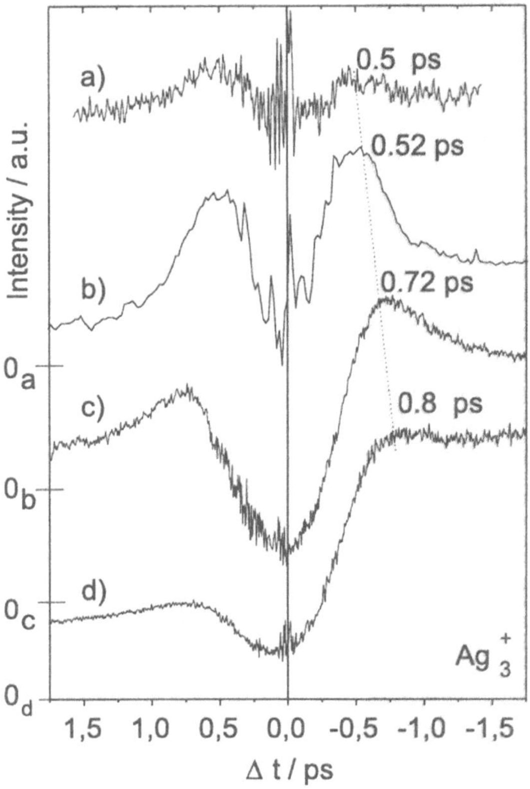

Fig. 5.3. Real-time spectra of the ultrafast structural relaxation of the silver trimer taken with wavelengths of **(a)** λ = 390 nm, **(b)** λ = 400 nm, **(c)** λ = 415 nm, **(d)** λ = 420 nm. Note that each curve has a different zero (0_a to 0_d), and that the time-independent background increases steadily with decreasing wavelength. The fine structure around $\Delta t = 0$ is, as in Fig. 2.23, due to interference of pump and probe pulses (taken from [421])

(Fig. 5.2 c). Note that this increase is not observed to the same degree for the fragment peak Ag_2^+. This indicates that the ionization process is less likely to lead to dissociation if there is some time for the relaxation of the neutral cluster.

In Fig. 5.3, the yield of Ag_3^+ is displayed as a function of the delay time Δt for various wavelengths of the detachment and ionization laser. At $\Delta t = 0$, the pump and probe laser pulses exchange their roles. The traces are not symmetrical, as the respective fluences of the pump and probe pulses are different. At long wavelengths (Fig. 5.3 c, d), the ion yield rises from almost zero to a maximum around $\Delta t = 720$ fs, and then decays to a constant value (saturation) at longer time delays. There it remains constant for more than 100 ps, the longest time delay used in the experiment performed here. This phenomenon is progressively washed out if light of shorter wavelength is used. In the case of 390 nm (Fig. 5.3 a) the ionization efficiency is almost independent of the delay time Δt. The time required to reach the maximum grows with increasing wavelength from 500 fs (Fig. 5.3 a) to about 800 fs (Fig. 5.3 d) consistent with the notion that the extra energy goes at least in part into the bending vibration. The yield of the charged cluster fragments Ag_2^+ and

Ag^+ exhibits the same delay time dependence as the trimer cation. From this, it can be concluded that the fragments were formed from Ag_3^+ after the multiphoton ionization. The signal at short times is more pronounced at relatively long wavelengths of the ionization laser, as then the ionization probability is strongly dependent on the vertical ionization potential in the momentary configuration of the neutral. The dependence of the cation yield on the power of the pump and probe laser pulses shows that the detachment process depends linearly, but the ionization process quadratically, on the respective light intensity. This is in good agreement with the creation process of the cations described above.

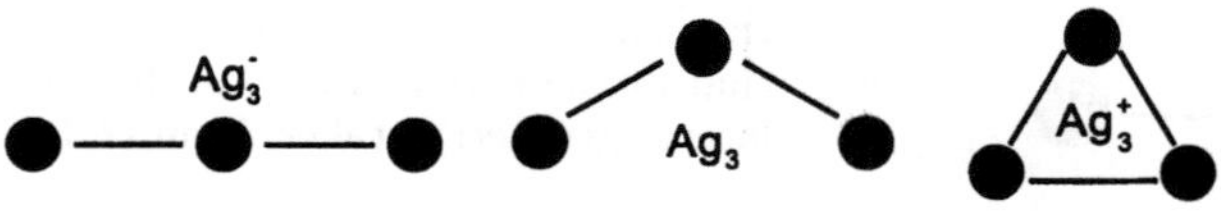

Fig. 5.4. The most stable geometry of the trimer is linear as a negative ion, obtuse isosceles as a neutral, and equilateral as a positive ion [84, 419]

Theoretical predictions indicate that the most stable geometry of the trimer is linear as a negative ion, obtuse isosceles as a neutral, and equilateral as a positive ion [84, 136, 419]. These geometries are sketched in Fig. 5.4. Insofar as the detachment and ionization process can be treated as a vertical Franck–Condon process, this can be expected to give rise to the following situation. The neutral trimer is presumably generated in a linear and therefore highly vibrationally excited configuration, at a saddle point, from which it bends slowly at first and then faster, passes through the geometry of the obtuse isosceles minimum, and then decelerates until it approaches equilateral geometry, where its overlap with the positive ion is greatest. In principle, any vibrations of the trimer excited in the photodetachment process may reveal themselves as recurrent peaks in the time-dependent photoionization probability, and be identified by frequency from the Fourier transform of the time-dependent, mass-analyzed photoion current [22]. If such signals can be found, they will constitute vibrational spectra of size-selected neutral clusters prepared in well-defined initial states comparable to the situation described in Chap. 3.

In fact, the results obtained with the silver trimers have not shown the multiply periodic behavior of a vibrational spectrum measured in the time domain. Rather, some of the results reveal less of the vibrational spectra but more of the dynamics of the internal rearrangements of these species, as shown by the time-dependent currents of positive ions in Fig. 5.3.

A first working interpretation of the behavior of the trimer is this: initially, the neutral is produced in a linear configuration by the vertical, Franck–Condon detachment process. Figure 5.5 summarizes neatly this hypothesis. The Franck–Condon overlap factor of the linear neutral with the equilateral

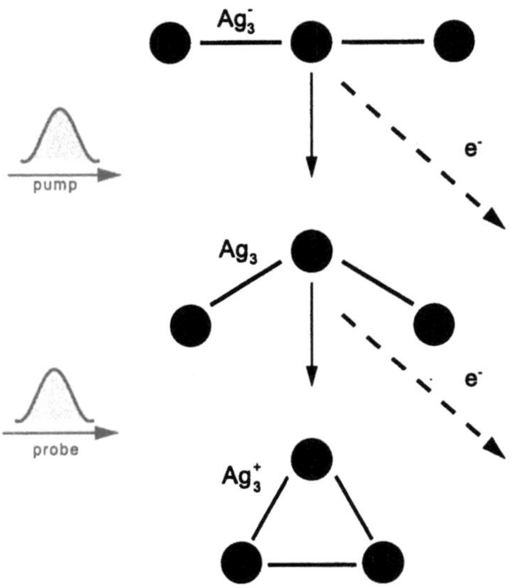

Fig. 5.5. Ultrafast structural relaxation of the silver trimer's molecular configuration starting in the anion's linear geometry. Photodetachment by the pump pulse initiates the bending of the meanwhile neutral trimer. The probe pulse can most efficiently produce cations while the neutral's configuration is close to that of the cation, which is the equilateral geometry (taken from [223])

positive ion is so low that virtually no positive ions are generated. However, the ultrashort photodetachment process puts the neutralized trimer in an extreme noneqilibrium situation (see Fig. 5.6). As a consequence of that, the structural relaxation process occurs. Hence, the neutral bends, passes through the obtuse equilibrium geometry of the neutral, and approaches the equilateral equilibrium geometry of the positive ion. During this time the positive-ion signal grows, and reaches a maximum when the system reaches its classical turning point near the equilateral triangular geometry. Then the system rebounds and the signal decreases. Figure 5.7 nicely follows this hypothesis. However, this model does not explain why the signal changes as a function of the laser pulse energy.

Combining a microscopic electronic theory with molecular dynamics simulations in the Born–Oppenheimer approximation, Bennemann, Garcia, and Jeschke presented the first theoretical results for the ultrafast structural changes in the silver trimer [135]. They determined the timescale for the relaxation from the linear to a triangular structure initiated by a photodetachment process and showed that the time-dependent change of the ionization potential (IP) reflects in detail the internal degrees of freedom.

In Fig. 5.8 the calculated real-time spectra of the fraction of silver trimers $p(h\nu, t)$ with IPs $V_{\rm i}(t) \leq h\nu$ are shown [135]. The sum of the pump and probe laser pulse energies $h\nu$, is scaled with $V_{\rm i}^0$ which is the minimal IP of Ag_3 in the equilateral equilibrium geometry of Ag_3^+. The quantity $p(h\nu, t)$ can be interpreted as the real-time probability for ionization with energy $h\nu$, and can be regarded as proportional to the real-time spectra obtained in the experiment described in this section. As in the experimental results, slightly above the silver trimer's IP the trimer can only be ionized after a certain

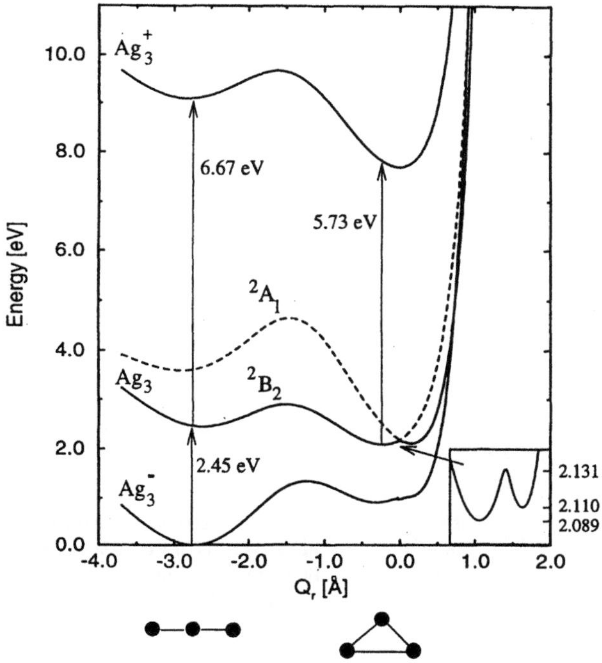

Fig. 5.6. One-dimensional cuts of the PESs of the ground states of Ag_3^-, Ag_3, and Ag_3^+ along $Q_r = -(Q_x^2) + Q_y^2)^{1/2}$ for fixed values of the polar angle $\alpha = \arctan(Q_x/Q_y) = 120°$ and $Q_s = 2.81$ Å(by courtesy of M. Hartmann, taken from [136]). Q_s, Q_x, and Q_y correspond to the symmetric stretching, the bending, and the antisymmetric stretching coordinate of the Ag_3^+ cation, respectively. The chosen Q_s value is that of the neutral's equilibrium nuclear configuration. The vertical electron detachment energy is 2.45 eV, the vertical ionization energy for the linear transition state and the equilibrium geometry of the neutral are 6.67 eV and 5.73 eV, respectively. The *dashed line* is the first excited electronic state of Ag_3

delay time $\Delta t(h\nu) \approx 750$ fs. After a maximum is reached, there is also a saturation of the signal, which remains constant.

Jeschke, Garcia, and Bennemann [134, 135] interpret their results as follows. Upon photodetachment of a binding electron, vibrational excitations occur, in particular those of the central atom along the chain direction. Owing to the shape of the PES the motion of the central atom dominates the real-time response over the first few hundred femtoseconds. Then, the slower, thermally activated bending motion comes into play and yields triangularly bonded Ag_3. The resultant bond formation is exothermic. The excess energy can in turn cause bond breaking or, in the case of uniform energy distribution, also a regular vibrational mode such as pseudorotation.

The vibrational excitation is high enough for the excited molecular modes to mix, and after the rebound, the still-unionized neutral trimers are left with enough energy to, for example, pseudorotate through their three equivalent

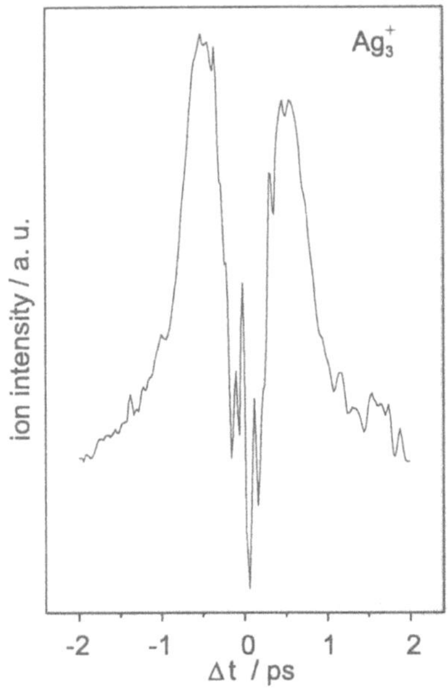

Fig. 5.7. Real-time structural relaxation of the neutral silver trimer after excitation with a 100 fs laser pulse of wavelength 400 nm(taken from [223])

obtuse-triangular equilibrium structures, going around the trough of their 'Mexican hat' potential energy surface [367, 387, 422, 423]. In so doing, they remain at a roughly constant distance from the equilateral geometry. Hence, the Franck–Condon factor also remains nearly constant, and therefore so does the positive-ion signal. However, this would mean that the pseudorotation has an extremely long mean life, which seems rather improbable. If the pseudorotation had a long lifetime the autocorrelation would show an oscillatory behavior. As shown by Bennemann and coworkers [134, 135], these correlations are strongly damped. They find the pseudorotation's lifetime to be slightly lower than the duration of one cycle. This demonstrates that pseudorotations do not play an important role in the dynamics of the silver trimer. The high kinetic energy of the trimer's atoms seems to prevent the occurrence of a regular mode such as pseudorotation. Hence, the saturation of the signal can, rather, be interpreted as a statistical effect induced by the temperature [135].

Furthermore, there might seem to be a possible inconsistency between the observation of Ag_2^+ made here and the energies reported in [419]. If the energy of the dissociation $Ag_3^+ \rightarrow Ag_2^+ + Ag$ is about 2.9 eV and the ionization potential of Ag_3 is approximately 5.7 eV, as [419] gives, then the dissociative ionization could only occur from vibrationally excited molecules. Of course the neutrals produced by photodetachment are expected to be quite vibrationally excited, since they should be nearly linear, but the energy deficit of 2.66 eV seems to be too large to make this explanation plausible. Simultaneous three-photon absorption has an energy deficit of over 2 eV and no

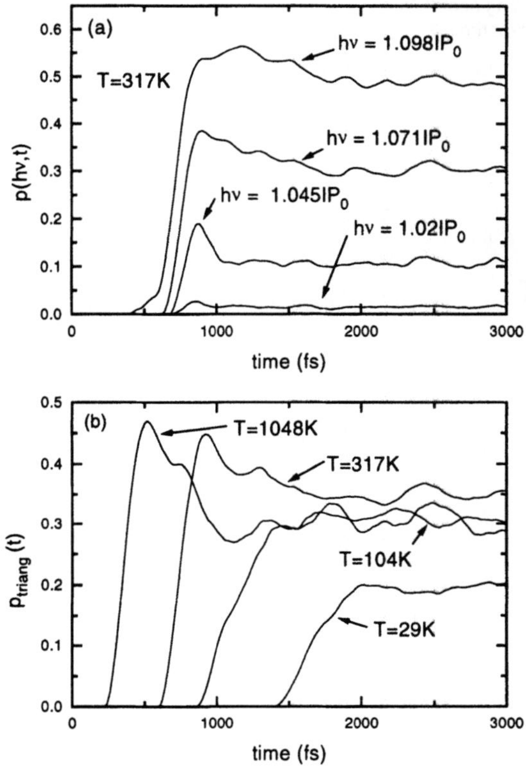

Fig. 5.8. (a) Real-time evolution of the fraction of clusters $p(h\nu, t)$ with $V_i(t) \leq h\nu$. The sharp increase and the overall time dependence of $p(h\nu, t)$ for increasing $h\nu$ should be compared with the experimental results presented in Fig. 5.3. The initial temperature of the trimer was $T = 317\,\text{K}$. (b) Real-time evolution of the fraction of clusters $p_{\text{triang}}(t)$ having triangular structures, for different initial temperatures but constant $h\nu$ (by courtesy of M. Garcia; taken from [135])

significant reservoir of vibrational energy, since the negative ions are relatively cool. Hence, it must be suspected that the fragment ions of Fig. 5.2 a, b are due to four-photon absorption.

Another approach has been chosen by Bonačić-Koutecký and coworkers [136]. Together with Jortner, they applied the density matrix method in the Wigner representation (see Sect. 2.2.3) to investigate the vibrational dynamics of the trimer. Within this approach the simulation of the real-time photoion spectra involves three steps:

- generating a classical phase space density of the Ag_3 anion in its ground state at a given temperature,
- propagating this phase space density classically on the potential energy surface of the ground electronic state of the neutral Ag_3, and
- calculating the real-time photoion signal.

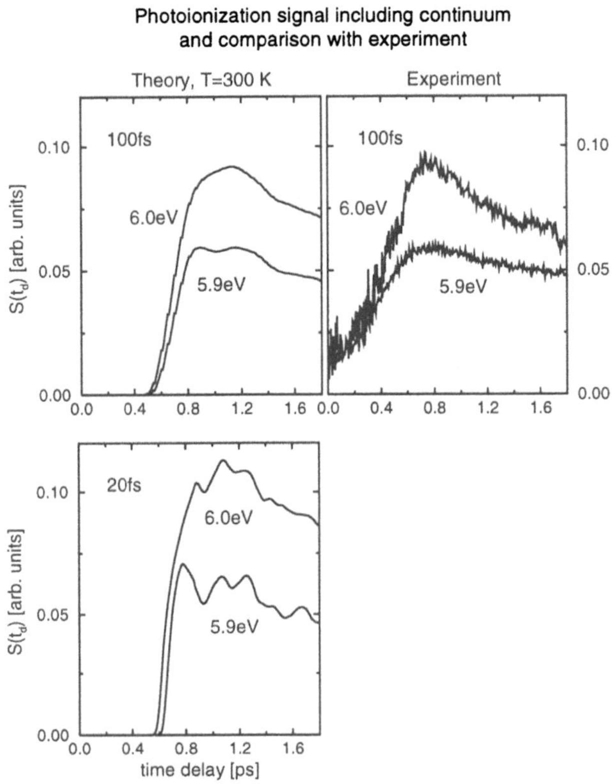

Fig. 5.9. Comparison of the real-time charge reversal process $Ag_3^- \rightarrow Ag_3^0 \rightarrow Ag_3^+$. **(a)** Simulation applying the density matrix method in the Wigner representation (T_{anions} = 300 K, 100 fs pulse width). **(b)** Pump&probe experiment (T_{anions} = 300 K, 100 fs pulse width). **(c)** Simulation applying the density matrix method in the Wigner representation in case of a 5 times shorter probe pulse width (by courtesy of M. Hartmann; taken from [136])

As is shown in Fig. 5.9, the calculated real-time photoion spectrum (Fig. 5.9 a) is in good agreement with the experimental results (Fig. 5.9 b). The increase of the photoion signal is finished after ~ 800 fs, which these authors also attribute to the geometrical relaxation of the formerly linear trimer. Bonačić-Koutecký states that the earlier onset of the increase is due to a direct three-photon ionization from the anion to the cation, which generates a broadening of the signal. While in Fig. 5.9 a a probe pulse of 100 fs FWHM is considered, in Fig. 5.9 c the result of a simulation performed with even shorter pulses reveals interesting features. The high temporal resolution enables the direct observation of coherent vibrations in the real-time spectrum. The analysis of the nuclear dynamics shows that these vibrations result from intramolecular vibrational redistribution of the symmetric bending mode into the symmetric stretch mode of the trimer.[2] Hence, for times greater than 1 ps the observed oscillations are a superposition of both vibrational excitations. For higher excitation energies the real-time spectra loose their structure and are temporally broadened. This can be attributed to the energy loss caused by the emitted electrons during the photoionization process.

5.2 Small Silver Clusters

In this section the first investigations of the ultrafast structural relaxation of larger silver clusters will be described. Comparison is made with the results obtained for Ag_3. The measurements were carried out using Ag_5^-, Ag_7^-, and Ag_9^- ions, which are produced with an intensity of approximately 100 pA, 50 pA, and 30 pA respectively. For the detachment pulse, wavelengths of 420 nm (Ag_5) and 400 nm (Ag_7 and Ag_9) were used, so that a one-photon detachment of the anions was possible in each case. The ionization was performed nonresonantly using two photons of the same wavelength. Owing to the alternating IP of Ag_n clusters and the available wavelength regime, therefore, only clusters with odd numbers of silver atoms were studied.

The geometries of the most stable cation, neutral, and anion configurations of the pentamer and 9-mer are drawn in Fig. 5.10 [84, 419]. The pentamer is expected to have a planar trapezoidal negative ion (with point group symmetry C_{2v}), a trigonal bipyramidal positive ion, and a neutral with nearly degenerate minima at both of these structures. The heptamer is expected to have pentagonal bipyramidal global mimima for all three charge states; Ag_9 is expected to be a trans-bicapped pentagonal bipyramid for its negative ion, and, for neutral and cation, an assembly of six tetrahedra with shared faces and C_{2v} symmetry [84, 419]. Insofar as the detachment and ionization processes can be treated as vertical Franck–Condon processes, these can be expected to give rise to several situations.

The pentamer negative ion has high overlap with one of the two lowest minima of the neutral, so its photodetachment has a favorable Franck–Condon factor, but the cross section for photoionization to a low-lying region of the PES of the positive ion becomes large only if the cluster isomerizes to its trigonal bipyramidal structure. The barrier for this isomerization is not yet known but may be evaluated theoretically in the near future. The heptamer should be relatively easy to photodetach and ionize without much change of geometry, and the 9-mer should have a vertical detachment energy about 0.4 to 0.5 eV above the adiabatic detachment energy of approximately 1.93 eV, with the difference available for the vibrational excitation associated

[2] This IVR process is induced by an intracluster collision, which corresponds to the strong repulsion of the terminal atoms at their closest approach. The collision manifests itself in a sharp reflection of the trimer from a motion across the bending coordinate to a motion along the stretching mode.

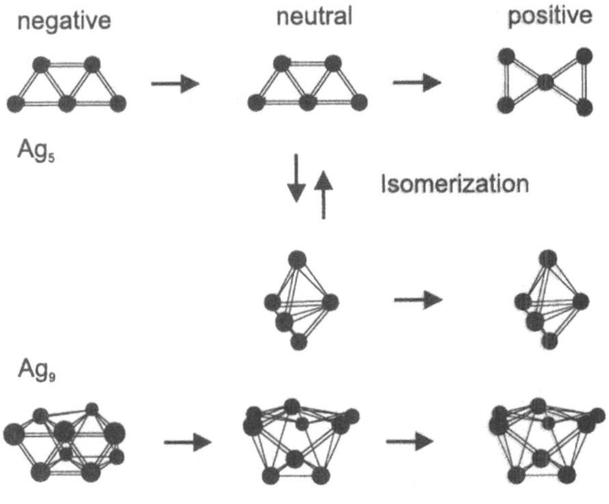

Fig. 5.10. Geometries of the most stable cation, neutral, and anion configurations of pentamer and 9-mer [84, 419]

with the relatively small change in geometry from the capped rhomboidal prism of the negative ion to the C_{2v}-symmetry structure of the neutral and positive clusters.

Before discussing the time dependence of the charge reversal process, it is necessary to have a short look at the fragmentation behavior of the concerned species. Figure 5.11 presents mass spectra of the cations ($Ag_{n=3,5,9}$) extracted from the trap after applying detachment and ionization pulses with zero time delay. Similar results are found for the heptamer. In all cases, most of the clusters stay intact during the detachment as well as the ionization process. Nevertheless a certain percentage of fragment ions is detectable, and the amount of fragmentation increases with cluster size.

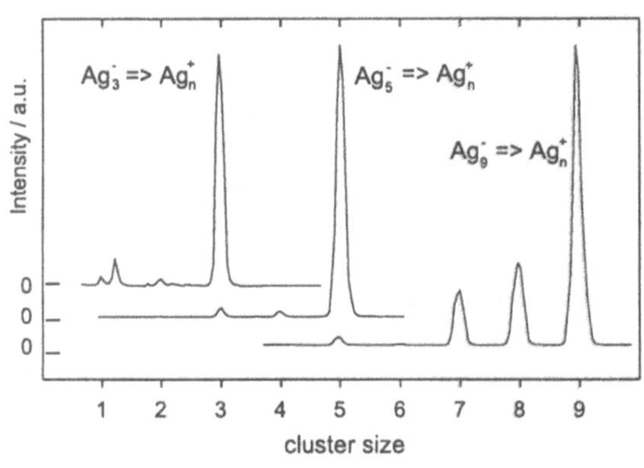

Fig. 5.11. Mass spectra of the cationic products from the process $Ag_n^- \longrightarrow Ag_n \longrightarrow Ag_m^+$ for $n = 3$, $n = 5$, and $n = 9$ (taken from[424])

As mentioned in Sect. 5.1, in the neutral silver trimer there is no excited electronic state which could be reached with a wavelength of 400 nm, and therefore the ionization is due to a nonresonant two-photon absorption process. The absorption spectra of the larger clusters are not known, but from the low cation intensity found in the experiment, it can be concluded that the $Ag^+_{n=3,5,9}$ ions are also produced by a nonresonant ionization of the corresponding neutrals. This again demonstrates that with ultrashort laser pulses, nonresonant ionization with little fragmentation is indeed possible.

Next, the real-time spectra $I(\Delta t)$ of the various cations extracted from the trap were monitored. For all cluster sizes ($Ag_{n=3,5,7,9}$) the fragment ions showed the same dependence on the delay time as the parent ions. Thus, it can be assumed that the fragmentation takes place after the ionization, and that the clusters survive the detachment intact. Regarding the dynamics of the neutral cluster, it is therefore sufficient to concentrate on the unbroken ions.

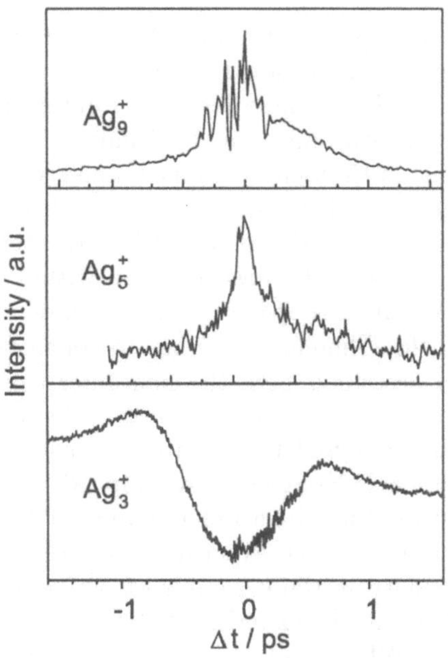

Fig. 5.12. Real-time mapping of the ultrafast structural response of $Ag_{n=3,5,9}$ to the charge reversal process (taken from [383])

The temporal evolution of the ion yield provides information on the vibrational dynamics in the neutral ground state. Figure 5.12 shows the time-resolved yield of $Ag^+_{n=5,9}$ compared to that of Ag_3. The curves for Ag_5 and Ag_9 exhibit a pronounced peak at $\Delta t = 0$ fs and a gradual descent to a constant value for longer delay times Δt. The decay time of this decent is 667 fs for Ag_9 and 283 fs for Ag_5. The data were obtained by deconvoluting the real-time spectra with the overall system response of the measuring system. The FWHM of the system response was estimated by a least-squares fit procedure to be close to 200 fs (see Fig. 5.13). The real-time response of the heptamer is rather weak; the decay time constant is close to 500 fs. As shown in Sect. 5.1, for Ag_3 the intensity rises to a maximum at $\Delta t = 720\,\mathrm{fs}$, and then decays to a finite level within 1.25 ps.

The irregular structure in all three spectra around $\Delta t = 0$ is due to the interference of the pump and the probe pulse; however, it is not resolved in these experiments as seen, for example, in Fig. 2.23. It has been checked that the ion signal depends linearly on the power of the pump and quadratically on the power of the probe laser, supporting our interpretation of successive one-photon detachment and two-photon ionization. The observed traces do depend on the wavelength used, but the overall features as described above are the same for all wavelengths used so far. Taking into account the known structures of the silver clusters (see Fig. 5.10), from these results it can be assumed that after a first relaxation the excess energy is rapidly redistributed among the degrees of freedom of the cluster. Therefore, the ion signal rapidly converges to a constant value.

The case of the silver pentamer deserves special attention. While for the anion a planar trapezoidal structure is by far the most stable one, the cation is a trigonal bipyramid. The neutral has nearly degenerate energy minima at both of these structures. Hence, the neutral pentamer might initially be created as its planar isomer. If it then forms its three-dimensional isomer after some time, there would be a much larger Franck–Condon overlap with the anion structure, and a strong increase in the ion signal would be expected after this structural relaxation within a picosecond. However, even within a period of 100 ps the monitored pentamer real-time spectrum shows no deviation from a constant value of the ion yield. Therefore, one has to conclude that under the experimental conditions described here the pentamer cannot overcome the barrier between its two lowest-energy structures and stays planar. Currently, experiments are being conducted to activate this isomerization process thermally. Warken and Bonačić-Koutecký have applied their theoretical approach (see Sect. 2.2.3) to study the relaxation dynamics of Ag_5 [309]. The results are not directly comparable to the experimental findings, since for their investigations they assumed an initial ensemble of Ag_5^- clusters at zero temperature, compared to 300 K in the experiment. Besides this, they used δ pulses instead of the 100 fs (FWHM) pulses used in the experiment. By calculating first the nuclear potential, followed by an analysis of the stationary properties of Ag_5, they identified important mode-coupling schemes of the neutral state. With the relevant Franck–Condon factors determined, they finally performed the time propagation of the neutral's state prepared by the detachment process. A pronounced energy flow to a bending motion was observed on a picosecond time-scale, which might be compared to the relaxation within $\approx 300\,\mathrm{fs}$ observed in the experiment. The deviation might, on the one hand, be caused by temperature effects but might, on the other hand, be partly due to the different pulse widths applied to the charge reversal process in the theory and the experiment.

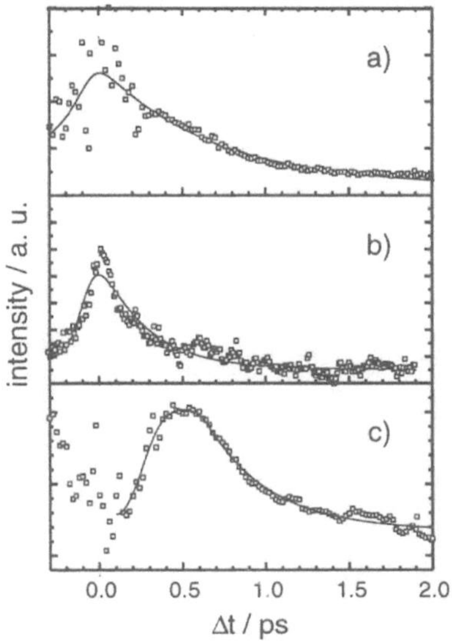

Fig. 5.13. Temporal evolution of the NeNePo signal of silver clusters Ag_n taken with $\lambda = 400\,\text{nm}$. **(a)** $n = 9$, **(b)** $n = 5$, **(c)** $n = 3$. The solid lines were given by a simple exponential decay model which was convoluted with the system response time in **(a)** and **(b)**, and a smooth interpolation in **(c)**. The fine structure around $\Delta t = 0$ arises from the interference between the pump and the probe pulse (taken from [424])

Summarizing the results, a strongly different behavior between Ag_3 on one side and $Ag_{n=5,7,9}$ on the other is observed. In the latter case, the variation of the ion signal is low, and not much of the underlying dynamics has been revealed (up to now!). This might be due to the fact that the probability of a nonresonant TPI does not strongly depend on the momentary geometry of the neutral. The reason for this might be that the energy of the two photons (detachment and ionization) is always sufficiently above the vertical ionization potential. This situation could be improved if a resonant ionization pathway via an intermediate excited state of the neutral could be used. In this case, the ionization probability depends strongly on the Franck–Condon overlap between the ground state and the excited state. Then, the ionization signal would be much more sensitive to the momentary configuration. The use of the tw-OPA, for example, (see Sect. 2.1.2) might open the doorway to exploit the known resonances in these clusters.

6. Summary and Forward View

In this book an overview has been given of the amazing opportunities provided by femtosecond real-time spectroscopy applied to small molecules and clusters. Fascinating phenomena such as control of molecular dynamics, selective state preparation, observation of vibrational wave packets on ground state PESs, ultrafast IVR, and photodissociation with unexpected and sometimes exceptional features have been introduced.

A summary of the investigations described in the preceding chapters is given in Sect. 6.1. In Sect. 6.2 we look forward to a few of the next future challenges and new goals of this modern and highly vital branch of natural sciences.

6.1 Summary

The main purpose of this book was to introduce the exciting field of femtosecond real-time spectroscopy applied to small model molecules and clusters. To carry out investigations in this field first of all an experimental setup was assembled with the following characteristics:

- high-repetition-rate laser pulse sources (80 MHz and 1 kHz) with sub-100 fs pulse duration (optionally 1 to 1.5 ps) and extremely low pulse-to-pulse fluctuations, covering the spectral region from $\sim 1\,\mu\text{m}$ to ~ 300 nm
- pump&probe setup to record the evolution in time of ion signals with a time resolution down to less than 1 fs
- stable, high-density beam source of extremely cold molecules/clusters for long-time studies (several hours)
- sensitive detection of ions with sufficient mass resolution ($m/\Delta m > 200$).

These experimental conditions enabled the detection of real-time MPI spectra with an excellent signal-to-noise ratio, allowing convincing Fourier analysis with a resolution of better than $0.1\,\text{cm}^{-1}$.

The time evolution of the MPI signal showing the ultrafast dynamics of the excited molecules and clusters was examined for several model systems, such as the sodium and potassium dimers and trimers. Different wave packet propagation phenomena could be studied in detail. Great importance was attached to the control of molecular dynamics, as well as to the effect of perturbation on the wave packet propagation. The potassium trimer presents a surprisingly limiting case, in which wave packet propagation on a repulsive PES is found.

The characteristics of the real-time MPI spectra drastically change, both for higher excitation energy (C and D state of Na_3) and for larger-size sodium and potassium aggregates. To observe the vibrational dynamics of larger molecules and clusters, the new 'NeNePo' technique was used. This technique, applied to silver trimers, convincingly shows the power of this experimental method because it allows the direct observation of the ultrafast change of the molecular geometry during the NeNePo process. Structural relaxation times can be determined.

The following sections give a brief summary of the results obtained.

6.1.1 Wave Packet Propagation in Alkali Dimers

In the investigations of wave packet propagation in the potassium dimer excited to its $A\,^1\Sigma_u^+$ state, it was demonstrated that this spectroscopic technique is a highly sensitive method of studying excited vibrational states and their perturbation due to crossing electronic states, and of controlling the molecular dynamics by the intensity of the exciting laser pulse. The high-repetition-rate laser pulse source and the excellent stability of the molecular beam machine enabled the detection of wave packet motion in real time for more than 200 ps. Besides straightforward Fourier analysis, time-windowed Fourier transform power spectra – spectrograms – are introduced to directly reveal the frequency content of the real-time spectra as a function of the time delay between pump and probe pulses.

First, the $^{39,39}K_2$ A state was studied in a spectral region ($\lambda = 840\,nm$) where no perturbation by superimposed electronic states appears. A fast oscillation with a period of 500 fs was observable, reflecting wave packet propagation in the excited A state. The time-resolved 3PI spectra were compared to quantum dynamical calculations of the wave packet propagation and showed excellent agreement. The transition mechanism, a pure (1+2)-photon process, can be analyzed. The two-photon ionization step is located at the outer turning point of the prepared wave packet on the A state PES. This allows the selective detection of the 500 fs vibration of the A state.

With a change of the excitation wavelength ($\lambda = 833.7\,nm$), the influence of the crossing $b\,^3\Pi_u$ state could be investigated. The highly mass-resolved detection allowed the isotope-selective ($^{39,39}K_2$ and $^{39,41}K_2$) exploration of the wave packet motion for more than 200 ps. Hence, in a direct comparison, the varying strength of the perturbation and its effect on the wave packet propagation could be studied for both isotopes. For $^{39,39}K_2$, a pronounced beat structure with a period of 10 ps is superimposed on the fast 500 fs oscillation. Fourier analysis of the 200 ps scan enabled a detailed identification of the excited vibronic levels of the A state with a resolution of better than $0.1\,cm^{-1}$. Energy shifts of the levels $v = 12$ ($\Delta E = 1.2\,cm^{-1}$) and $v = 13$

($\Delta E = 2.1\,\mathrm{cm}^{-1}$), due to spin–orbit coupling with the crossing b state, were estimated. For $^{39,41}K_2$, however, under identical excitation conditions, practically no perturbation was noticeable. The pronounced change of the real-time spectra found for these closely related systems neatly demonstrates the high sensitivity of this experimental technique to study level shifts of vibrational states, with a resolution of better than $0.1\,\mathrm{cm}^{-1}$. Theoretical simulations of the real-time spectra on the basis of fully quantum dynamical calculations reproduce well the experimental results for both isotopes.

Increasing the intensity of the pump pulse ($\lambda = 833.7\,\mathrm{nm}$) by a factor ~ 10 opens the challenging field of controlled molecular dynamics. It was possible to prove that the laser intensity can be used for a defined 'femtosecond state preparation' and as a consequence to control the molecular dynamics. In the recorded real-time ion signal the effects of two processes could be distinguished. For moderate laser intensities the MPI process is as previously described. For high intensities, however, resonant impulsive stimulated Raman scattering (RISRS) appears. The real-time ion signal obtained then reflects the induced dynamics of the K_2 ground state. The high ionization probability is located at the inner turning point of the ground state PES. A three-photon process is necessary, subsequently, to ionize the prepared dimer from the ground state.

The influence of perturbing electronic states on wave packet propagation was also demonstrated for the Na_2 $A\,^1\Sigma_u^+$ state, which is closely related to that of the K_2 A system. Changing the wavelength so as to excite in a spectral region with (642 nm) and without (620 nm) perturbation, the revival structure of the real-time spectra is completely different. A Fourier analysis revealed the reason for this drastic change. Comparing the spectra with RKR calculations, a shift of the A state's vibrational level $v = 8$ of $\Delta E = 0.85\,\mathrm{cm}^{-1}$ was determined. The shift is – as in the case of K_2 – caused by spin–orbit coupling of the crossing A and $b\,^3\Pi_u$ states. This again neatly highlights the high sensitivity of femtosecond real-time spectroscopy. In addition, the transition pathways for this 3PI process were deduced directly from the oscillating pattern of the transient ion signal. While for the unperturbed case the ionization takes place at the inner turning point of the A state PES, for the perturbed case it occurs at the outer turning point. The $(2)\,^1\Pi_g$ state in both cases acts as the Franck–Condon window.

Similar results were obtained for two isotopes of the heteronuclear NaK molecule excited to its electronic A state. By means of spectrogram technique, the pronounced revival structure of the induced wave packet propagation was nicely illustrated. Higher order fractional revivals could be observed for the first time for a molecular system. Strong half and quarter revivals were observed at $T_{\mathrm{frev}}^{1,2(23,39)} \cong 75\,\mathrm{ps}$ and $T_{\mathrm{frev}}^{1,4(23,39)} \cong 40\,\mathrm{ps}$ for the lighter isotope $^{23}Na^{39}K$. For the heavier but rare isotope $^{23}Na^{41}K$ the respective times were $T_{\mathrm{frev}}^{1,2(23,41)} \cong 68\,\mathrm{ps}$ and $T_{\mathrm{frev}}^{1,4(23,41)} \cong 35\,\mathrm{ps}$. Weak fractional revivals could be estimated in the case of $^{23}Na^{39}K$ from the spectrograms of the real-time data, at $T_{\text{frev}}^{1,3(23,39)} \approx 25\,\text{ps}$ and $T_{\text{frev}}^{2,3(23,39)} \approx 50\,\text{ps}$. The advantages of three-photon ionization spectroscopy of extremely cold molecules compared to fluorescence detection techniques of molecules in a heat-pipe oven have been shown.

6.1.2 Wave Packet Propagation in Alkali Trimers

For the bound Na_3 B system, 1.25 ps and 120 fs one-color real-time TPI spectroscopy, at moderate intensities, yielded preferential excitations of the relatively slow pseudorotation (3 ps) and the fast symmetric stretch mode (310 fs) respectively. Three-dimensional quantum chemical and quantum dynamical ab initio investigations fully corresponded to these experimental results. The time-dependent wave packet dynamics elucidate the effect of ultrafast state preparation on the molecular dynamics. Hence, these experiments manifest efficient control of molecular dynamics using the pulse duration as a control parameter. Since known cw spectra show the pseudorotation features only, this result demonstrates also that cw and femtosecond spectroscopy have complementary sensitivities for the excitation of different vibrational modes, thus neatly confirming the original conjecture of Zewail (see e.g. [425]).

The 3PI spectra of K_3 excited at 800 nm presented a suprisingly limiting case. Wave packet propagation superimposed on a fast decay of the ion intensity within a few picoseconds was observable. Conventional cw spectroscopy failed to detect an electronic state predicted by theoretical calculations. This state is expected to show analogous features to the well-known B state of Na_3. The Fourier spectrum reveals three frequencies (66, 82, and $109\,\text{cm}^{-1}$), which can be assigned to the symmetric and asymmetric bending modes of the excited state and the symmetric stretch mode of the K_3 ground state respectively. No pseudorotation dynamics were found for the excitation wavelengths used. Similar results were obtained when the trimer was excited with $\lambda = 820\,\text{nm}$. The decay of the ion signal, described by a single exponential decay, is caused by ultrafast photodissociation. The decay constants vary, depending on the excitation wavelength, between 4 and 7 ps. This result again demonstrates the complementarity of cw and femtosecond real-time spectroscopy. Only femtosecond real-time spectroscopy, using very intense laser pulses with a broad spectrum, can open a temporal window to detect this 3PI process.

6.1.3 Ultrafast Photodissociation in Small Alkali Clusters

The ultrafast photodissociation dynamics of the Na_3 C state was analyzed with time-resolved two-color TPI spectroscopy in the picosecond regime. The two excitation wavelengths required, independently tunable for the pump and the probe pulse, were generated by a home-built synchronization of two mode-locked titanium sapphire lasers. The deconvoluted real-time spectra can be well described by a single exponential decay with a time constant strongly dependent on the excited vibrational bands. Starting from the lower-lying vibrational bands, the decay time decreases from 1.12 ns to 12 ps, providing detailed information about the onset of the C state's predissociation.

The application of real-time MPI spectroscopy to electronic transitions of higher excitation energy or of larger aggregates reveals the rapidly growing number of different dissociation channels. To study the photodissociation dynamics of extremely cold sodium and potassium clusters, femtosecond two-color pump&probe spectroscopy was used. An energy-level model was developed to describe the total photodissociation dynamics in a cluster beam consisting of a mixture of cluster sizes. The model enables, by means of a least-squares-fit algorithm, the filtering out of the pure fragmentation dynamics of a specified cluster size under well-defined excitation conditions. For the D state of Na_3, the fragmentation time constants are in the region of 260 to 640 fs, depending on the excitation wavelength. Lifetimes of photoexcited larger sodium clusters were also estimated. Depending on cluster size and excitation wavelength, they are between 200 fs and 900 fs. Different fragmentation behaviors were found for odd- and even-numbered clusters. Odd-numbered sodium clusters excited with $\sim$ 840 nm radiation tended to dissociate more rapidly.

For potassium clusters $K_{n=3...9}$, fragmentation time constants in the range of 200 fs to a few picoseconds, depending on excitation energy and cluster size, were found by applying the same energy-level model. No even–odd alternations, but a monotonic increase of the photoionization probability up to K_8 followed by a sudden decrease, were found for excitation at 2.00 eV and 2.94 eV. This particular instability for K_8 is contrary to the expectations from jellium model considerations. The expected higher stability, however, was indeed found when exciting K_8 at 1.47 eV. Owing to the excited repulsive PES, wave packet propagation either for Na_n or for K_n shows up under these excitation conditions.

6.1.4 Ultrafast Structural Relaxation in Small Silver Clusters

To overcome the reported ultrafast photodissociation of small clusters, a new experimental approach, called 'NeNePo', was developed. The technique is a quite general scheme for exploring the time evolution of a coherent nonequilibrium state in mass-selected neutral clusters. By photodetachment of Ag_n anions with sub-100 fs laser pulses ($\sim$ 400 nm), an ensemble of neutral Ag_n in the electronic ground state was generated. The real-time spectrum is a measure of the evolving Franck–Condon factor for photoionization of the neutral, prepared by a vertical photodetachment process from a low-lying vibrational state of the anion. It reveals the internal motion of the atoms which constitute the cluster, especially in cases in which detachment, ionization, or even both involve a significant change in geometry. For Ag_3, the change from the linear to the obtuse triangular configuration could be estimated to take place within 500 to 750 fs. An elegant theoretical description of the observed process was given by Bennemann and coworkers. Their theory, a combination of an electronic theory and molecular dynamics calculations, allows one to determine the ultrafast structural response of optically excited small clusters. The estimated real-time dependence of the ionization potential of Ag_3 upon photodetachment was in excellent agreement with the experimental results. The first results on larger silver clusters are promising and may allow detailed information on, for example,isomerization times of the neutral or cation system.

6.2 Forward View

When solving or understanding problems it is nearly always the case that new questions and goals arise – this is no exception. Hence, most of the future prospects touched upon below are closely connected to the results presented here. The success of all of these investigations depends essentially on the synergetic application of theoretical and experimental concepts.

In recent years the theory of coherent chemistry has shifted towards attempts to actively control molecular dynamics rather than 'just' to observe passively these ultrafast phenomena. These studies have created huge interest and much activity in the optical and chemical communities [426–428]. The first promising experiments have confirmed the fascinating theoretical predictions (see in [428]).

One of the main goals of coherent control is to actively govern reactions; this certainly belongs to the 'Holy Grails' of chemistry. Controlling a reaction by use of light means controlling the nature and products of a chemical reaction by, for example, breaking a certain selected bond in one of the molecules involved in the reaction. A nice prototype for controlling reactions in this way might be the 'simple' reaction of water with sodium. Although bulk sodium reacts strongly exothermically with water, a single sodium atom does not. This raises the question of the minimum size a sodium molecule must have to start the reaction

$$\mathrm{Na} + \mathrm{H_2O} \rightarrow \mathrm{NaOH} + (1/2)\mathrm{H_2}. \tag{6.1}$$

A promising candidate might be the solvated sodium dimer excited to its double-minimum state.[1] Employing a pump and control scheme, as shown in Fig. 6.1, will enable the transfer of the initially covalent bond-length isomer, to its ionic isomer which should favor the

$$\mathrm{Na_2(H_2O)}_n \rightarrow 2\mathrm{NaOH} + (n-2)\mathrm{H_2O} + \mathrm{H_2} \tag{6.2}$$

[1] Delacrétaz and Wöste [429] and Hertel et al. [430] have investigated the spectroscopy, and Baumert and Gerber [29] the real-time dynamics, of the double minimum of Na_2.

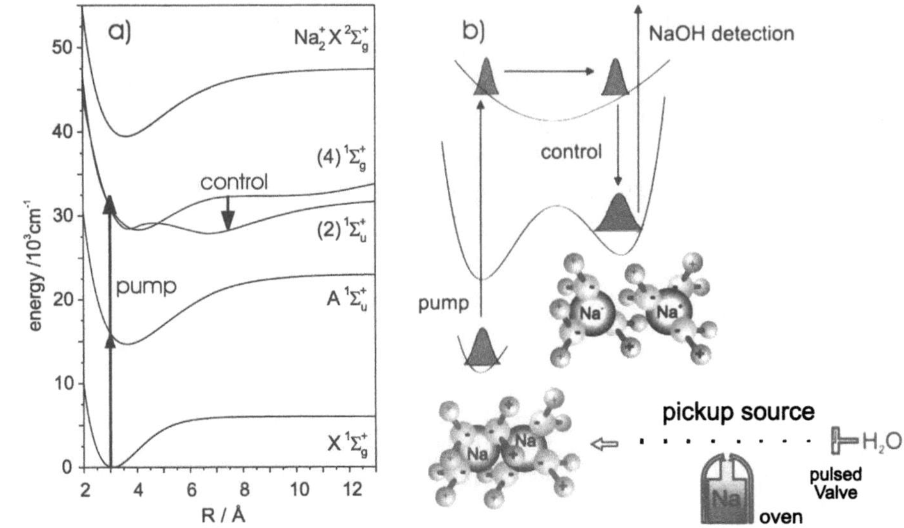

Fig. 6.1. Pump&control scheme for $Na_2(H_2O)_n$. (**a**) Potential-energy surfaces of Na_2 with double minimum state $(2)\,^1\Sigma_u^+$ and shelf state $(4)\,^1\Sigma_g^+$. (**b**) A pickup cluster source produces $Na_2(H_2O)_n$ clusters, which are excited by two pump pulses to the shelf state. After a certain delay time the control pulse transfers the wave packet to the outer minimum of the double minimum state. Here the reaction of Na_2 with the water molecules takes place

reaction of the metal chromophore with the H_2O molecules. A third laser pulse might be used to ionize the reaction product NaOH.

Closely related to the experiments presented here are some fascinating investigations on the Cs_2 molecule performed by Girard and coworkers [37–40]. These authors carried out a one-color coherent control experiment, applying two identical but time-delayed ultrashort laser pulses to prepare wave packet (quantum) interferences in the bound, excited electronic B state. The temporal evolution of this wave packet was probed by the second, time-delayed pulse, photoionizing the dimer. The interferogram obtained exhibits high-frequency oscillations corresponding to Ramsey fringes (at the Bohr frequency of the transition) modulated by a slow envelope, which reveals the oscillation of the vibrational wave packet's recurrences (see Fig. 6.2). By controlling the relative phase of the two preparing pulses it was possible to control coherently the destructive and constructive wave packet interferences. This enables the control of the MPI process, i.e. of whether or not ionization takes place.

This temporal coherent control technique allows the extraction of detailed information on the structure of the excited molecular states with a precision similar to that of high resolution Fourier transform spectroscopy. It provides an efficient way of controlling the creation of a wave packet in a bound state. A simultaneous excitation of two or even more excited states may enable independent control of the formation of two or several wave packets. Owing to the high sensitivity of this interferometric technique, Girard's approach [40] provides an excellent tool for probing perturbations that could effect the molecule, especially inbetween the interaction of the two phase shifted laser pulses. Such perturbations might be collisions with other molecules or ac–Stark shifts, induced by the interaction with an intense laser pulse.

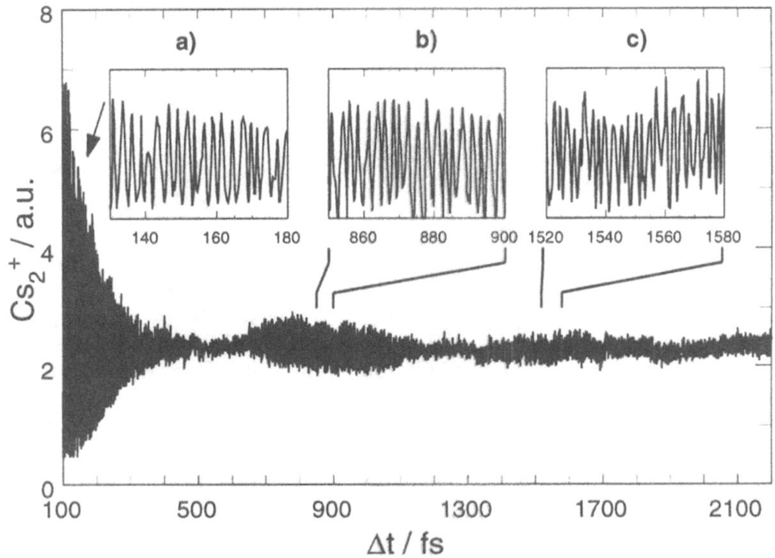

Fig. 6.2. Real-time interferogram of an MPI process in Cs_2 (by courtesy of B. Girard; taken from [40]). The dimer's B state was excited and subsequently ionized by a pair of 150 fs pump pulses to the B state (768 nm). The high-resolution (0.3 fs) scan was obtained with scanning Brewster plates. Insets show enlarged views **(a)** in the optical interference region, and of quantum interferences in the first **(b)** and second **(c)** vibrational recurrence. The fast oscillation has in all cases a period of 2.56 fs

Concerning the studies of perturbation, it is straightforward to look at heteronuclear systems. The lack of a center of symmetry widens the possibilities of interaction between different states of singlet and triplet multiplicities beyond the range of those possible in homonuclear dimers. In particular, the NaK molecule, forging a link between the well-understood Na_2 and K_2 molecules, was treated as a model system. Another possibility of breaking the symmetry of homonuclear dimers or trimers is to attach a hydrogen atom. The ligand can be considered as a weak perturbation and will allow the observation of its influence on the dynamics of, for example, intramolecular energy transfer processes. The Na_3 molecule, with its possibility of mode-selective excitation, is a preferred candidate for these explorations of intramolecular vibrational redistributions.

A great challenge in the femtosecond physics and chemistry community is to examine ultrafast isomerization processes. Here, small systems such as the triangular Na_2Li as well as the rhombic Na_2Li_2 or Na_2H_2 can be regarded as simple prototypes. Even larger complexes might be attached to the alkali molecules. Scoles and coworkers have used nanometer-sized helium droplets, each containing about 10^4 helium atoms, as an inert substrate on which they efficiently stuck alkali atoms [431, 432] and sodium trimers [433]. By laser-induced fluorescence they found the spectral perturbations to be limited. Femtosecond real-time spectroscopy should provide deeper insight into the dynamics of 'making and breaking' the bonding of this molecular cluster complex. Besides this, the 'shaking' and, especially, the intramolecular vibrational redistribution in an excited trimer is of great interest to the femtochemistry community. For example, the influence of the He cluster on the trimer's pseudorotation (see Sect. 3.2.1) might be an important challenge in the nearer future.

Further advances in the field of coherent chemistry will require improved time resolution and tunability of the laser sources in the experiments [434]. Shorter laser pulses with a pulse duration of less than 20 fs will be appropriate. The optimum control technique will be achieved by specially tailored femtosecond pulse shapes [435–439], determined by sophisticated calculations using the theories of coherent chemistry [440–444]. Finally, more elaborate polarization and multipulse excitation schemes, including chirped and ultrashort pulses, will be needed.

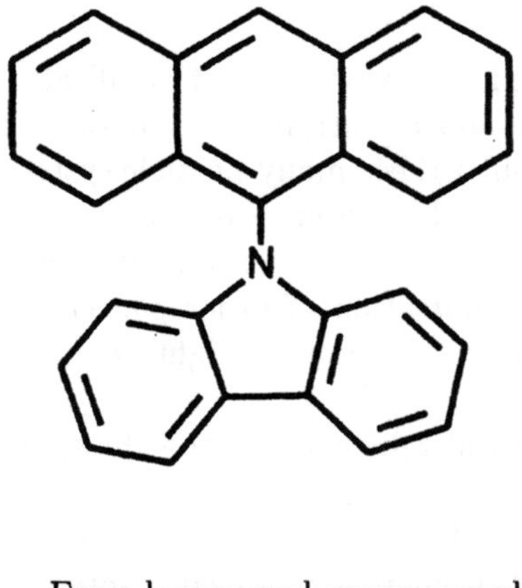

Fig. 6.3. Structure of the 9-(N-carbazolyl)-anthracene (C9A) molecule

Even larger and more complex molecules could be investigated by femtosecond real-time spectroscopy. Besides many others, the process of photoinduced twisting of molecular bonds is of fundamental importance in organic photochemistry. The 9-(N-carbazolyl)-anthracene (C9A) molecule, shown in Fig. 6.3, is an important model system for twisted intramolecular charge

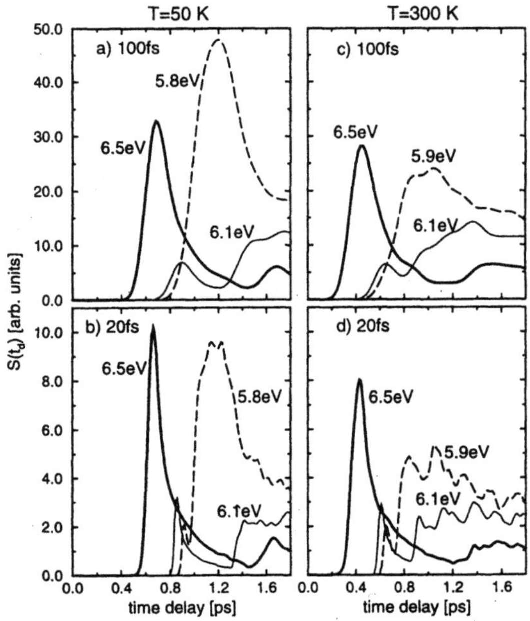

Fig. 6.4. Calculated real-time ZEKE photoionization spectrum of an ensemble of Ag_3 molecules at initial temperature 50 K. The asymmetry of the spectrum for larger delay times t between the pump and the probe pulse indicates the width of the initial distribution of the phase space density, as trajectories of lower initial velocity contribute to the signal later. The ultrafast relaxation from the linear to the obtuse triangle geometry of the trimer is visible (by courtesy of M. Hartmann; taken from [136]).

transfer (TICT). It has been treated in great detail theoretically by Manz and coworkers [445, 446] and experimentally by Rettig, Zimmermann, and coworkers [447, 448]. Simulated real-time MPI spectra on the torsional PES show that by tuning the wavelength of the pump pulse, the position of the PES crossing should be directly detectable in the structure of the ion signal. Here also, internuclear vibrational redistribution phenomena can be analyzed in great detail.

For additional studies of the ultrafast photodissociation of small alkali clusters, the support of detailed qualitative and quantitative predictions of the stability and photoinduced fragmentation of these many-particle systems is essential. Knowing that even for the three-body system Na_3, state-of-the-art time-dependent quantum dynamical simulations are extremely costly, requiring considerable computer time, the application of other concepts is necessary. In the near future, the density matrix formalism might be an appropriate approach here.

Further investigations on ultrafast structural relaxation of the charge reversal process will surely be carried out using different cluster systems and different ionization pathways. NeNePo might also be used to investigate reactive compounds (i.e. cluster ligand systems) where a chemical reaction or molecular rearrangement starts after the neutralization. The associated dynamics could eventually be deduced by detecting and energy-analyzing the photoelectrons of the probe process as a function of Δt.

Combining the promising NeNePo method with the zero-kinetic-energy (ZEKE) technique, as Jortner together with Bonačić-Koutecký and her group did with their theoretical approach [136], might provide detailed, highly time-resolved information on the nuclear dynamics of molecules and clusters. These authors studied the ZEKE photoionization spectrum for an ensemble of Ag_3 molecules at low temperatures and found the temporal resolution to be much higher than in 'normal' real-time charge reversal experiments. As is seen in Fig. 6.4, a pronounced structure is visible. Close to the Franck–Condon transition a sharp peak is observed in the ion signal, when using energies of about 6.5 eV. From the real-time change of the silver atom's probability density these authors conclude that the peak is due to the obtuse triangle geometry. The relaxation of the acute triangle geometry (see Fig. 5.4) continues as long as the Coulomb forces of the two silver atoms do not stop their approach. Within a very short period of time, nearly the complete energy of the bending mode is transferred to the symmetric stretch mode. Since the ionization potential is nearly constant (5.8 eV) along the stretching mode, the strong peak at 1.2 ps seems to be a consequence of this intramolecular vibrational redistribution process. These stimulating results can be regarded as an excellent basis for experimental investigations of ultrafast energy transfer during charge reversal processes, and not only in Ag_3.

Besides this the real-time observation of charge reversal processes might allow the direct view of the isomerization dynamics of special molecules. Vinylidene ($H_2C{=}C$), the simplest member of the class of unsaturated carbenes, can be regarded as a prototype for this. From photoelectron spectroscopy it is well known that the C_2H_2 anion has vinylidene structure[449]. In its natural ground state it rather rapidly becomes its more stable isomer acetylene ($HC{=}CH$). The NeNePo scheme should allow the estimation of at least the corresponding isomeriztion time.

Another great challenge is to extend the 'NeNePo' experiments to extremely cold molecules and clusters produced by adiabatic expansion. The femtosecond real-time spectroscopy of these cold, mass-selected molecules and clusters will permit characterization of the ground state of a great variety of aggregates with high precision. Na_3 and Li_3, with similar characteristic geometries to Ag_3 for the anion, neutral, and cation, might be promising candidates.

The Na_3 ground state, compared to the excited electronic states, is rather complicated. The lowest vibronic states are localized but, at higher energy, the PES is affected by the linear configuration and a second-order Jahn–Teller treatment is not possible. As a consequence, it is not straightforward to deduce the ground state PES from the known experimental results. The NeNePo approach, however, will give selective access to all vibronic states. Thus the crossing of the pseudorotation barrier, which corresponds to a drastic change from localized to Jahn–Teller pseudorotation states, can be analyzed in great detail. Photodetaching e.g. cold Li_3 will prepare a wave packet propagating towards the minima of the two almost degenerate bending vibrations, since Li_3 is unstable at the high-order saddle point of the linear configuration. Interferences varying with the vibronic level prepared in the ground state should then be detectable in the signal of this rather light molecule. Valuable information about high-order saddle points of the PES could be obtained.

Beyond straightforward studies of the real-time evolution of the neutral molecules and clusters, it should be possible to do things to the size-selected neutrals. They could be excited vibrationally or electronically by radiation, for example. Perhaps the most fascinating possibility of the 'NeNePo' method would be, to study in real-time, reactive collisions of size-selected molecules and clusters with substances added to the interaction region of the laser pulses and molecular beam.

However, with this kaleidoscopic run through a few prospects, I should like to finish this look at the possible near future, knowing that in real-time the realization will take billions of femtoseconds!

References

1. G. Porter, 'Flash Photolysis into the Femtosecond – a Race against Time' in *Femtosecond Chemistry*, J. Manz and L. Wöste (eds.) (VCH, Weinheim, 1995), Vol. 1, p 3.
2. *Femtochemistry: Ultrafast Dynamics of the Chemical Bond,* Vols. 1 & 2, *World Scientific Series in 20th Century Chemistry,* A.H. Zewail (ed.) (World Scientific, Singapore, 1994).
3. *Femtosecond Chemistry,* Vols. 1 & 2, J. Manz and L. Wöste (eds.) (VCH, Weinheim, 1995).
4. J. Manz and A.W. Castleman Jr., 'Femtosecond Chemistry' in *Femtosecond Chemistry,* Vol. 97 *Special Issue of the Journal of Physical Chemistry,* J. Manz and A.W. Castleman Jr. (eds.) (American Chemical Society, Washington, 1993), pp. 12 423.
5. *Femtochemistry: Ultrafast Chemical and Physical Processes in Molecular Systems,* M. Chergui (ed.) (World Scientific, Singapore, 1996).
6. P.W. Atkins, *Quanta* (Oxford University Press, Oxford, 1991).
7. J. Manz, 'Molecular Wave Packet Dynamics: Theory and Experiments 1926–1996' in *Femtochemistry and Femtobiology,* V. Sundström (ed.) (World Scientific, Singapore, 1998).
8. B.M. Garraway and K.-A. Suominen, 'Wave Packet Dynamics: New Physics and Chemistry in Femto-Time', Rep. Prog. Phys. **58**, 365 (1995).
9. S. Brandt and H.D. Dahmen, *The Picture Book of Quantum Mechanics* (Springer, New York, 1994).
10. V. Brückner, K.-H. Feller, and U.-W. Grummt, *Applications of Time-Resolved Optical Spectroscopy,* Vol. 66 *Studies in Physical and Theoretical Chemistry* (Elsevier, Amsterdam, 1990).
11. W.H. Knox, R.L. Fork, M.C. Downer, R.H. Stolen, C.V. Shank, and J.A. Valdmanis, 'Optical Pulse Compression to 8 fs at a 5-kHz Repetition Rate', Appl. Phys. Lett. **46**, 1120 (1985).
12. R.L. Fork, C.H. Brito Cruz, P.C. Becker, and C.V. Shank, 'Compression of Optical Pulses to Six Femtoseconds by Using Cubic Phase Compensation', Opt. Lett. **12**, 483 (1987).
13. *Ultrashort Laser Pulses and Applications,* Vol. 60 *Topics in Applied Physics,* W. Kaiser (ed.) (Springer, Berlin, Heidelberg, 1988), .
14. G.H.C. New, 'Femtofascination', Physics World **7**, 33 (1990).
15. M.J. Rosker, M. Dantus, and A.H. Zewail, 'Femtosecond Real-Time Probing of Reactions. I. The Technique', J. Chem. Phys. **89**, 6113 (1988).
16. M. Dantus, M.J. Rosker, and A.H. Zewail, 'Femtosecond Real-Time Probing of Reactions. II. The Dissociation Reaction of ICN', J. Chem. Phys. **89**, 6128 (1988).
17. L.R. Khundkar and A.H. Zewail, 'Ultrafast Molecular Reaction Dynamics in Real-Time: Progress Over a Decade', Ann. Rev. Phys. Chem. **41**, 15 (1990).

18. M. Dantus, M.H.M. Janssen, and A.H. Zewail, 'Femtosecond Probing of Molecular Dynamics by Mass-Spectrometry in a Molecular Beam', Chem. Phys. Lett. **181**, 281 (1991).
19. A.H. Zewail, 'Femtochemistry', J. Phys. Chem. **97**, 12 427 (1993).
20. A.H. Zewail, 'Laser Femtochemistry', Science **242**, 1645 (1988).
21. M. Dantus, R.M. Bowman, and A.H. Zewail, 'Femtosecond Laser Observations of Molecular Vibration and Rotation', Nature **343**, 737 (1990).
22. M. Gruebele, G. Roberts, M. Dantus, R.M. Bowman, and A.H. Zewail, 'Femtosecond Temporal Spectroscopy and Direct Inversion to the Potential: Application to Iodine', Chem. Phys. Lett. **166**, 459 (1990).
23. M. Gruebele and A.H. Zewail, 'Femtosecond Wavepacket Spectroscopy: Coherences, the Potential, and Structural Determination', J. Chem. Phys. **98**, 883 (1993).
24. I. Fischer, D.M. Villeneuve, M.J.J. Vrakking, and A. Stolow, 'Femtosecond Wave-Packet Dynamics Studied by Time-Resolved Zero-Kinetic Energy Photoelectron Spectroscopy', J. Chem. Phys. **102**, 5566 (1995).
25. M.J.J. Vrakking, I. Fischer, D.M. Villeneuve, and A. Stolow, 'Collisional Enhancement of Rydberg Lifetimes Observed in Vibrational Wave Packet Experiments', J. Chem. Phys. **103**, 4538 (1995).
26. T. Baumert, B. Bühler, R. Thalweiser, and G. Gerber, 'Femtosecond Spectroscopy of Molecular Autoionization and Fragmentation', Phys. Rev. Lett. **64**, 733 (1990).
27. T. Baumert, B. Bühler, M. Grosser, R. Thalweiser, V. Weiss, E. Wiedenmann, and G. Gerber, 'Femtosecond Time-Resolved Wave Packet Motion in Molecular Multiphoton Ionization and Fragmentation', J. Phys. Chem. **95**, 8103 (1991).
28. T. Baumert, M. Grosser, R. Thalweiser, and G. Gerber, 'Femtosecond Time-Resolved Molecular Multiphoton Ionization: The Na_2 System', Phys. Rev. Lett. **67**, 3753 (1991).
29. T. Baumert and G. Gerber, 'Fundamental Interactions of Molecules (Na_2, Na_3) with Intense Femtosecond Laser Pulses', Isr. J. Chem. **34**, 103 (1994).
30. T. Baumert, R. Thalweiser, V. Weiss, and G. Gerber, 'Femtosecond Time-Resolved Photochemistry of Molecules and Metal Clusters' in *Femtosecond Chemistry*, J. Manz and L. Wöste (eds.) (VCH, Weinheim, 1995), Vol. 2, p 397.
31. A. Assion, T. Baumert, V. Seyfried, V. Weiss, E. Wiedemann, and G. Gerber, 'Femtosecond Spectroscopy of the (2) $^1\Sigma_u^+$ Double Minimum State of Na_2: Time Domain and Frequency Spectroscopy', Z. Phys. D. **36**, 265 (1996).
32. T. Baumert, V. Engel, C. Röttgermann, W.T. Strunz, and G. Gerber, 'Femtosecond Pump–Probe Study of the Spreading and Recurrence of a Vibrational Wave Packet in Na_2', Chem. Phys. Lett. **191**, 639 (1992).
33. T. Baumert, V. Engel, C. Meyer, and G. Gerber, 'High Laser Field Effects in Multiphoton Ionization of Na_2. Experiment and Quantum Calculations', Chem. Phys. Lett. **200**, 488 (1992).
34. V. Engel, T. Baumert, Ch. Meier, and G. Gerber, 'Femtosecond Time-Resolved Molecular Multiphoton Ionization and Fragmentation of Na_2: Experiment and Quantum Mechanical Calculations', Z. Phys. D **28**, 37 (1993).
35. C. Meier and V. Engel, 'Electron Kinetic Energy Distributions from Multiphoton Ionization of Na_2 with Femtosecond Laser Pulses', Chem. Phys. Lett. **212**, 691 (1993).
36. Ch. Meier and V. Engel, 'Pump–Probe Ionization Spectroscopy of a Diatomic Molecule: The Sodium Molecule as a Prototype Example' in *Femtosecond Chemistry*, J. Manz and L. Wöste (eds.) (VCH, Weinheim, 1995), Vol. 1, Chap. 11, p 369.

37. V. Blanchet, M.A. Bouchene, O. Cabrol, and B. Girard, 'One-Color Coherent Control in Cs_2. Observation of 2.7 fs Beats in the Ionization Signal', Chem. Phys. Lett. **233**, 491 (1995).
38. V. Blanchet, M.A. Bouchène, and B. Girard, 'Femtosecond Time-Resolved Spectroscopy and Coherent Control in Cs_2' in *Fast Elementary Processes in Chemical and Biological Systems*, Vol. 364 *AIP Conference Proceedings*, A. Tramer (ed.) (AIP Press, Woodbury, New York, 1996), p 619.
39. V. Blanchet, M.A. Bouchène, and B. Girard, 'Coherent Control and Molecular Interferometry with Ultrashort Laser Pulses in Cs_2' in *Femtochemistry: Ultrafast Chemical and Physical Processes in Molecular Systems*, M. Chergui (ed.) (World Scientific, Singapore, 1996).
40. V. Blanchet, M.A. Bouchène, and B. Girard, 'Temporal Coherent Control in the Photoionization of Cs_2: Theory and Experiment', J. Chem. Phys. (submitted).
41. J.M. Papanikolas, R.M. Williams, P. Kleiber, J.L. Hart, C. Brink, S.D. Price, and S.R. Leone, 'Wave-Packet Dynamics in the $Li_2(^1\Sigma_g^+)$ Shelf State: Simultaneous Observation of Vibrational and Rotational Recurrences with Single Rovibronic Control of an Intermediate State', J. Chem. Phys. **103**, 7269 (1995).
42. R. de Vivie-Riedle, B. Reischl, S. Rutz, and E. Schreiber, 'Femtosecond Study of Multiphoton Ionization Processes in K_2 at Moderate Laser Intensities', J. Phys. Chem. **99**, 16 829 (1995).
43. S. Rutz, R. de Vivie-Riedle, and E. Schreiber, 'Femtosecond Wave Packet Propagation in Spin–Orbit Coupled Electronic States of $^{39,39}K_2$ and $^{39,41}K_2$', Phys. Rev. A **54**, 306 (1996).
44. S. Rutz, E. Schreiber, and L. Wöste, 'Wave Packet Propagation in Excited $^{39,39}K_2$ and $^{39,41}K_2$' in *Fast Elementary Processes in Chemical and Biological Systems*, Vol. 364 *AIP Conference Proceedings*, A. Tramer (ed.) (AIP Press, Woodbury, New York, 1996), p 652.
45. E. Schreiber, S. Rutz, and R. de Vivie-Riedle, 'Intensity Effects on Ionization Pathways in K_2: Control of the Wave Packet Dynamics' in *Laser in Forschung und Technik – Laser in Research and Engineering*, W. Waidelich, H. Hügel, H. Opower, H. Tiziani, R. Wallenstein, and W. Zinth (eds.) (Springer, Berlin, Heidelberg, 1996), pp. 203–212.
46. E. Schreiber, S. Rutz, and L. Wöste, 'Intensity Controlled Molecular Dynamics of the Potassium Dimer' in *Fast Elementary Processes in Chemical and Biological Systems*, Vol. 364 *AIP Conference Proceedings*, A. Tramer (ed.) (AIP Press, Woodbury, New York, 1996), p 645.
47. S. Rutz, E. Schreiber, and L. Wöste, 'Femtosecond Pump&Probe Spectroscopy on the K_2 Molecule. Perturbations in Different Isotopomeres' in *Ultrafast Processes in Spectroscopy*, O. Svelto, D. De Silvestri, and G. Denardo (eds.) (Plenum, New York, 1996), pp. 127–131.
48. S. Rutz, E. Schreiber, and L. Wöste, 'Femtosecond Vibrational Dynamics of the Potassium Dimer', Surf. Rev. Lett. **3**, 475 (1996).
49. S. Rutz and E. Schreiber, 'Real-Time Vibrational Dynamics of K_2' in *Ultrafast Phenomena IX*, Vol. 60 *Springer Series in Chemical Physics*, P.F. Barbara, W.H. Knox, G.A. Mourou, and A.H. Zewail (eds.) (Springer, Berlin, Heidelberg, 1994), p 312.
50. E. Schreiber, 'Wavepacket Propagation Phenomena in Small Molecules Induced by Ultrashort Laser Pulses' in *Proceedings of the International Conference on LASERS '95*, V.J. Corcoran and T. Goldman (eds.) (Society for Optical and Quantum Electronics, McLean (VA), 1996), p 53.
51. M.J.J. Vrakking, D.M. Villeneuve, and A. Stolow, 'Observation of Fractional Revivals of a Molecular Wave Packet', Phys. Rev. A **54**, R37 (1996).

52. S. Rutz and E. Schreiber, 'Fractional Revivals of Wave Packets in the $A^1\Sigma_u^+$ State of K_2. A Comparison of Two Different Pump&Probe Cycles by Spectrograms', Chem. Phys. Lett. **269**, 9 (1997).
53. J. Heufelder, H. Ruppe, S. Rutz, E. Schreiber, and L. Wöste, 'Fractional Revivals of Vibrational Wave Packets in the NaK $A^1\Sigma^+$ State', Chem. Phys. Lett. **269**, 1 (1997).
54. I.Sh. Averbukh and N.F. Perel'man, 'Fractional Revivals: Universality in the Long-Term Evolution of Quantum Wavepackets Beyond the Correspondence Principle Dynamics', Phys. Lett. A **139**, 449 (1989).
55. G. Stock and W. Domcke, 'Model Studies on the Time-Resolved Measurement of Excited-State Vibrational Dynamics and Vibronic Coupling', Chem. Phys. **124**, 227 (1988).
56. G. Stock and W. Domcke, 'Femtosecond Spectroscopy of Ultrafast Nonadiabatic Excited-State Dynamics on the Basis of *Ab-Initio* Potential-Energy Surfaces: The S_2 State of Pyrazine', J. Phys. Chem. **97**, 12466 (1993).
57. L. Seidner and W. Domcke, 'Microscopic Modelling of Photoisomerization and Internal-Conversion Dynamics', Chem. Phys. **186**, 27 (1994).
58. C. Daniel, M.-C. Heitz, J. Manz, and C. Ribbing, 'Spin–Orbit Induced Radiationless Transitions in Organometallics: Quantum Simulation of the $^1E \rightarrow {}^3A_1$ Intersystem Crossing Process in $HCo(CO)_4$', J. Chem. Phys. **102**, 905 (1995).
59. P. Brumer and M. Shapiro, 'Control of Unimolecular Reactions Using Coherent Light', Chem. Phys. Lett. **126**, 541 (1986).
60. D.J. Tannor and S.A. Rice, 'Control of Selectivity of Chemical Reactions via Wave Packet Evolution', J. Chem. Phys. **83**, 5013 (1985).
61. D.J. Tannor, R. Kosloff, and S.A. Rice, 'Coherent Pulse Sequence Induced Control of Selectivity of Reactions: Exact Quantum Mechanical Calculations', J. Chem. Phys. **85**, 5805 (1986).
62. B. Reischl, 'Quantum Dynamical Three-Dimensional *Ab-Initio* Approach to a Femtosecond Pump–Probe Ionization Spectrum of Na_3 (B) at Low Laser Field Intensities', Chem. Phys. Lett. **239**, 173 (1995).
63. B. Kohler, J.L. Krause, F. Raski, K.R. Wilson, and V.V. Yakovlev, 'Quantum Control and Experimental Realities', Acc. Chem. Res. **28**, 133 (1995).
64. B. Kohler, J.L. Krause, F. Raski, C. Rose-Petruck, R.M. Whitnett, K.R. Wilson, V.V. Yakovlev, and Y.J. Yan, 'Femtosecond Pulse Shaping for Molecular Control' in *Femtosecond Reaction Dynamics*, D.A. Wiersma (ed.) (North-Holland, Amsterdam, 1994), p 209.
65. D. Goswami, C.W. Hillegas, J.X. Tull, and W.S. Waren, 'Generation of Shaped Femtosecond Laser Pulses: New Appoaches to Laser Selective Chemistry' in *Femtosecond Reaction Dynamics*, D.A. Wiersma (ed.) (North-Holland, Amsterdam, 1994), p 291.
66. A.H. Zewail, 'Femtochemistry: Concepts and Applications' in *Femtosecond Chemistry*, J. Manz and L. Wöste (eds.) (VCH, Weinheim, 1995), Vol. 1, p 15.
67. U. Banin, A. Bartana, S. Ruhman, and R. Kosloff, 'Impulsive Excitation of Coherent Vibrational Motion Ground Surface Dynamics Induced by Intense Short Pulses', J. Chem. Phys. **101**, 8461 (1994).
68. G. Delacrétaz, E.R. Grant, R.L. Whetten, L. Wöste, and J. Zwanziger, 'Fractional Quantization of Molecular Pseudorotation in Na_3', Phys. Rev. Lett. **56**, 2598 (1986).
69. S. Rakowsky, R.F.W. Herrmann, and W.E. Ernst, 'High Resolution Laser Spectroscopy of the Na_3 B – X System', Z. Phys. D **26**, 273 (1993).
70. W.E. Ernst and S. Rakowsky, 'Is the B State of Na_3 a Case of Berry's Phase?', Z. Phys. D **26**, 270 (1993).

71. T. Baumert, R. Thalweiser, V. Weiß, and G. Gerber, 'Time-Resolved Studies of Neutral and Ionized Na_n Clusters with Femtosecond Light Pulses', Z. Phys. D **26**, 131 (1993).
72. V. Bonačić-Koutecký, P. Fantucci, and J. Koutecký, 'Theoretical Interpretation of the Photoelectron Detachment Spectra of Na^-_{2-5} and of Absorption Spectra of Na_3, Na_4, and Na_8 Clusters', J. Chem. Phys. **93**, 3802 (1990).
73. J. Gaus, *Strukturelle und elektronische Eigenschaften kleiner reiner, gemischter und dotierter Alkalimetall-Cluster*, Ph.D. thesis, Freie Universität Berlin, Berlin-Dahlem, 1995.
74. F. Cocchini, T.H. Upton, and W. Andreoni, 'Excited States and Jahn–Teller Interactions in the Sodium Trimer', J. Chem. Phys. **88**, 6068 (1988).
75. R. Meiswinkel and H. Köppel, 'A Pseudo-Jahn–Teller Treatment of the Pseudorotational Spectrum of Na_3', Chem. Phys. **144**, 117 (1990).
76. J. Schön and H. Köppel, 'Femtosecond Time-Resolved Ionization Spectroscopy of Na_3 (B) and the Question of the Geometric Phase', Chem. Phys. Lett. **231**, 55 (1994).
77. R. Meiswinkel and H. Köppel, 'A Pseudo-Jahn–Teller Treatment of the B System of Na_3', Z. Phys. D **19**, 63 (1991).
78. H. Köppel and R. Meiswinkel, 'Point-Charge Model for Vibronic Coupling Constants of Metal Atom Trimers', Z. Phys. D **32**, 153 (1994).
79. A.J. Dobbyn and J.M. Hutson, 'The Influence of the Ionisation Potential on the Simulated Ion Signal from Femtosecond Pump–Probe Spectroscopy', Chem. Phys. Lett. **236**, 547 (1995).
80. A.J. Dobbyn, *Chemical Dynamics Using Wave Packet Methods*, Ph.D. thesis, University of Durham, Durham, UK, 1993.
81. B. Reischl, *Quantentheorie zur Schwingungsstruktur und -dynamik des pseudorotierenden Na_3 (B)*, Ph.D. thesis, Freie Universität Berlin, Berlin-Dahlem, 1995.
82. V. Bonačić-Koutecký and J. Gaus, private communications.
83. L. Wöste, private communications.
84. V. Bonačić-Koutecký, L. Češpiva, P. Fantucci, J. Pittner, and J. Koutecký, 'Effective Core Potential-Configuration Interaction Study of Electronic Structure and Geometry of Small Anionic Ag_n Clusters: Predictions and Interpretation of Photodetachment Spectra', J. Chem. Phys. **100**, 490 (1994).
85. H. Haberland (ed.), *Clusters of Atoms and Molecules*, Vol. 52 *Springer Series in Chemical Physics*, (Springer, Berlin, Heidelberg, 1994).
86. H. Haberland (ed.), *Clusters of Atoms and Molecules II*, Vol. 56 *Springer Series in Chemical Physics*, (Springer, Berlin, Heidelberg, 1994).
87. *Structures and Dynamics of Clusters*, T. Kondow, K. Kaya, and A. Terasaki (eds.) (Universal Academy Press, Tokyo, 1996).
88. W. de Heer, 'The Physics of Simple Metal Clusters: Experimental Aspects and Simple Models', Rev. Mod. Phys. **65**, 611 (1993).
89. V. Bonačić-Koutecký, P. Fantucci, and J. Koutecký, 'Quantum Chemistry of Small Clusters of Elements of Groups Ia, Ib, and IIa: Fundamental Concepts, Predictions, and Interpretation of Experiments', Chem. Rev. **91**, 1035 (1991).
90. M. Kappes and S. Leutwyler, 'Spectroscopic Detection Methods' in *Atomic and Molecular Beam Methods*, G. Scoles (ed.) (Oxford University Press, New York, 1988), Vol. 1, Chap. 15, p 380.
91. *Elemental and Molecular Clusters*, Vol. 6 *Springer Series in Material Science*, G. Benedek, T.P. Martin, and G. Pacchioni (eds.) (Springer, Berlin, Heidelberg, 1987), .
92. H. Haberland, 'Cluster' in *Bergmann Schaefer*, Vol. 5 *Lehrbuch der Experimentalphysik*, W. Raith (ed.) (Walter de Gruyter, Berlin, 1992), p 550.

93. J. Jortner, 'Cluster Size Effects', Z. Phys. D **24**, 247 (1992).
94. N.D. Spencer and G.A. Somorjai, 'Catalysis', Rep. Prog. Phys. **46**, 1 (1983).
95. M.R. Zakin, R.O. Brickman, D.M. Cox, and A. Kaldor, 'Dependence of Metal Cluster Reaction Kinetics on Charge State. II. Chemisorption of Hydrogen by Neutral and Positively Charged Iron Clusters', J. Chem. Phys. **88**, 6605 (1985).
96. P. Fayet, F. Granzer, G. Hegenbart, E. Moisar, B. Pischel, and L. Wöste, 'Latent-Image Generation by Deposition of Monodisperse Silver Clusters', Phys. Rev. Lett. **55**, 3002 (1985).
97. G. Delacrétaz, P. Fayet, J.P. Wolf, and L. Wöste, 'Spectroscopy, Reactivity, and Photodynamics of Size-Selected Metal Clusters' in *Proceedings of the International School of Physics "Enrico Fermi", Course CVII*, G. Scoles (ed.) (North-Holland, Amsterdam, 1990), pp. 359–396.
98. M.A. Duncan and D.H Rouvray, 'Microclusters', Scientific American (Dec.), 60 (1989).
99. N. Lee, R.G. Keessee, and A.W. Castleman Jr., 'On the Correlation of Total and Partial Enthalpies of Ion Solvation and the Relationship to the Energy Barrier to Nucleation', J. Colloid Interface Sci. **75**, 555 (1980).
100. C.P. Schulz and I.V. Hertel, 'Atoms in Polar Solvents' in *Clusters of Atoms and Molecules*, Vol. 56 *Springer Series in Chemical Physics*, H. Haberland (ed.) (Springer, Berlin, Heidelberg, 1994), p 7.
101. N.F. Scherer, L.R. Khundkar, R.B. Bernstein, and A.H. Zewail, 'Real-Time Picosecond Clocking of the Collision Complex in a Bimolecular Reaction: The Birth of HO from $H+CO_2$', J. Chem. Phys. **87**, 1451 (1987).
102. T. Halicioglu and C.W. Bauschlicher Jr., 'Physics of Microclusters', Rep. Prog. Phys. **51**, 883 (1988).
103. J. Jortner, 'Cluster Size Effects Revisited', J. Chim. Phys. **92**, 205 (1995).
104. D. Scharf, J. Jortner, and U. Landmann, 'Excited-State Dynamics of Rare-Gas Clusters', J. Chem. Phys. **88**, 4273 (1988).
105. A. Amirav, U. Even, and J. Jortner, 'Electronic–Vibrational Excitations of Aromatic Molecules in Large Argon Clusters', J. Phys. Chem. **86**, 3345 (1982).
106. L. Bewig, U. Buck, C. Mehlmann, and M. Winter, 'Ionization Induced Fragmentation of Size Selected Neutral Sodium Clusters', J. Chem. Phys. **100**, 2765 (1994).
107. I.V. Hertel, C.P. Schulz, A. Goerke, H. Palm, and G. Leipelt, 'Fragmentation Analysis of Size Selected Sodium Clusters', J. Chem. Phys. **107**, 3528 (1997).
108. J. Tiggesbäuker, L. Köller, H.O. Lutz, and K.H. Meiwes-Broer, 'Giant Resonances in Silver-Cluster Photofragmentation', Chem. Phys. Lett. **190**, 42 (1992).
109. C. Bréchignac, Ph. Cahusac, J. Leygnier, and J. Weiner, 'Dynamics of Unimolecular Dissociation of Sodium Cluster Ions', J. Chem. Phys. **90**, 1492 (1989).
110. C. Bréchignac, Ph. Cahusac, R. Pflaum, and J.-Ph. Roux, 'Adiabatic Unimolecular Dissociation of Heterogeneous Alkali Clusters', J. Chem. Phys. **88**, 3732 (1988).
111. C. Bréchignac, Ph. Cahusac, J.-Ph. Roux, D. Pavolini, and F. Spiegelmann, 'Adiabatic Decomposition of Mass-Selected Alkali Clusters', J. Chem. Phys. **87**, 5694 (1987).
112. V. Bonačić-Koutecký, J. Pittner, C. Scheuch, M.F. Guest, and J. Koutecký, 'Quantum Molecular Interpretation of the Absorption Spectra of Na_5, Na_6, and Na_7 Clusters', J. Chem. Phys. **96**, 7938 (1992).
113. V. Bonačić-Koutecký, P. Fantucci, C. Fuchs, C. Gatti, J. Pittner, and S. Polezzo, '*Ab initio* Predictions of Optically Allowed Transitions in Na_{20}.

Nature of Exitations and Influence of Geometry', Chem. Phys. Lett. **213**, 522 (1993).

114. A. Herrmann, M. Hofmann, S. Leutwyler, E. Schumacher, and L. Wöste, 'Optical Spectroscopy of Na_3 by Two-Photon-Ionisation in a Supersonic Molecular Beam', Chem. Phys. Lett. **62**, 216 (1979).
115. L. Wöste, *Massenselektive Laserspektroskopie an Metallclustern in Überschallmolekularstrahlen*, Ph.D. thesis, Universität Bern, Bern, 1978.
116. F.H. Kühling, *Ultrakurzzeitspektroskopie an kleinen Alkaliclustern*, Ph.D. thesis, Freie Universität Berlin, Berlin-Dahlem, 1993.
117. J.L. Martins, J. Buttet, and R. Car, 'Electronic and Structural Properties of Sodium Cluster', Phys. Rev. B **31**, 1804 (1985).
118. *Proceedings of the ISSPIC 5 (Konstanz 1990), Fifth International Symposium on Small Particles and Inorganic Clusters* (Z. Phys. D, No. 1–4, 1990), Vol. 19, 20.
119. *Proceedings of the ISSPIC 6 (Chicago 1992), Sixth International Symposium on Small Particles and Inorganic Clusters*, I.V. Hertel (ed.) (Z. Phys. D, No.1–4, 1993), Vol. 26, 27.
120. *Proceedings of the ISSPIC 7 (Kobe, Japan 1994), Seventh International Symposium on Small Particles and Inorganic Clusters*, Y. Nishina and S. Sugano (eds.) (Surf. Rev. Lett., Vol. 3, 1996), No. 1.
121. *Proceedings of the ISSPIC 8 (Copenhagen 1996), Eighth International Symposium on Small Particles and Inorganic Clusters*, I.V. Hertel (ed.) (Z. Phys. D, No.1-4, 1997), Vol. 40.
122. R. Schinke, *Photodissociation Dynamics, Spectroscopy and Fragmentation of Small Polyatomic Molecules, Cambridge Monographs on Atomic, Molecular and Chemical Physics* (Cambridge University Press, Cambridge, 1993).
123. D.M. Willberg, M. Gutmann, J.J. Breen, and A.H. Zewail, 'Real-Time Dynamics of Clusters. I. I_2X_n $(n = 1)$', J. Chem. Phys. **96**, 198 (1992).
124. M. Broyer, G. Delacrétaz, P. Labastie, J.P. Wolf, and L. Wöste, 'Size-Selective Depletion Spectroscopy of Predissociated States of Na_3', Phys. Rev. Lett. **57**, 1851 (1986).
125. J.X. Wang, P.D. Kleiber, K.M. Sando, and W.C. Stwalley, 'Fine-Structure Branching in the Near-Threshold Photodissociation of NaK $(X^1\Sigma^+ - B^1\Pi)$', Phys. Rev. A **42**, 42 (1990).
126. C.R. Wang, S. Pollak, D. Cameron, and M.M. Kappes, 'Optical Absorption Spectroscopy of Sodium Clusters as Measured by Collinear Molecular Beam Photodepletion', J. Chem. Phys. **93**, 3787 (1990).
127. C. Wang, S. Pollak, T.A. Dahlseid, G.M. Koretsky, and M.M. Kappes, 'Photodepletion Probes of Na_5, Na_6, and Na_7. Molecular Dimensionality Transition (2D→3D)?', J. Chem. Phys. **96**, 7931 (1992).
128. A.H. Zewail and R.B. Bernstein, 'Real-Time Laser Femtochemistry: Viewing the Transition from Reagents to Products', Chem. & Engin. **66**, 24 (1988).
129. A.H. Zewail and R.B. Bernstein, 'Real-Time Laser Femtochemistry: Viewing the Transition from Reagents to Products' in *The Chemical Bond, Structure and Dynamics*, A.H. Zewail (ed.) (Academic Press, Boston, 1992), p 223.
130. A.H. Zewail, 'Femtosecond Transition-State Dynamics', Faraday Discuss. Chem. Soc. **91**, 207 (1991).
131. T. Baumert, R. Thalweiser, and G. Gerber, 'Femtosecond Two-Photon Ionization Spectroscopy of the B-State of Na_3-Clusters', Chem. Phys. Lett. **209**, 29 (1993).
132. T. Baumert, R. Thalweiser, V. Weiß, and G. Gerber, 'Femtosecond Dynamics of Molecular and Cluster Ionization and Fragmentation' in *Ultrafast Phenomena VIII*, Vol. 55 *Springer Series in Chemical Physics*, J.-L. Martin, E.P. Ippen,

G.A. Mourou, and A.H. Zewail (eds.) (Springer, Berlin, Heidelberg, 1993), p 83.

133. G. Herzberg, *Molecular Spectra and Molecular Structure* (Krieger, Malabar, Florida, 1991), Vol. III.
134. H.O. Jeschke, M.E. Garcia, and K.H. Bennemann, 'Analysis of the Ultrafast Dynamics of the Silver Trimer upon Photodetachment', J. Phys. B: At. Mol. Opt. Phys. **29**, L545 (1996).
135. H.O. Jeschke, M.E. Garcia, and K.H. Bennemann, 'Theory for the Ultrafast Structural Response of Optically Excited Small Clusters: Time Dependence of the Ionization Potential', Phys. Rev. A **54**, R4601 (1996).
136. M. Hartmann, J. Pittner, V. Bonačić-Koutecký, A. Heidenreich, and J. Jortner, **108**, No.8, Feb. 22nd (1998).
137. S.L. Shapiro (ed.), *Ultrashort Light Pulses, Picosecond Techniques and Applications*, Vol. 18 *Topics in Applied Physics*, (Springer, Berlin, Heidelberg, 1977).
138. C.V. Shank, E.P. Ippen, and S.L. Shapiro (eds.), *Picosecond Phenomena*, Vol. 4 *Springer Series in Chemical Physics*, (Springer, Berlin, Heidelberg, 1978).
139. R.M. Hochstrasser, W. Kaiser, and C.V. Shank (eds.), *Picosecond Phenomena II*, Vol. 14 *Springer Series in Chemical Physics*, (Springer, Berlin, Heidelberg, 1980).
140. K.B. Eisenthal, R.M. Hochstrasser, W. Kaiser, and A. Laubereau (eds.), *Picosecond Phenomena III*, Vol. 23 *Springer Series in Chemical Physics*, (Springer, Berlin, Heidelberg, 1982).
141. D.H. Auston and K.B. Eisenthal (eds.), *Ultrafast Phenomena IV*, Vol. 38 *Springer Series in Chemical Physics*, (Springer, Berlin, Heidelberg, 1984).
142. G.R. Flemming and A.E. Siegmann (eds.), *Ultrafast Phenomena V*, Vol. 46 *Springer Series in Chemical Physics*, (Springer, Berlin, Heidelberg, 1986).
143. J. Herrmann and B. Wilhelmi, *Lasers for Ultrashort Laser Pulses* (Akademie Verlag, Berlin, 1987).
144. R.J.H. Clark and R.E. Hester, *Advances in Spectroscopy* (Wiley, Chichester, 1989), Vol. 18.
145. T. Yajima, K. Yoshihara, C.B. Harris, and S. Shionoya (eds.), *Ultrafast Phenomena VI*, Vol. 48 *Springer Series in Chemical Physics*, (Springer, Berlin, Heidelberg, 1988).
146. E. Klose and B. Wilhelmi (eds.), *Ultrafast Phenomena in Spectroscopy*, Vol. 49 *Springer Proceedings in Physics*, (Springer, Berlin, Heidelberg, 1990).
147. C.B. Harris, E.P. Ippen, G.A. Mourou, and A.H. Zewail (eds.), *Ultrafast Phenomena VII*, Vol. 53 *Springer Series in Chemical Physics*, (Springer, Berlin, Heidelberg, 1990).
148. J.-L. Martin, A. Migus, G.A. Mourou, and A.H. Zewail (eds.), *Ultrafast Phenomena VIII*, Vol. 55 *Springer Series in Chemical Physics*, (Springer Verlag, Berlin, Heidelberg, 1993).
149. *Ultrafast Dynamics of Chemical Systems*, Vol. 7 *Chemical Reactivity*, J.D. Simon (ed.) (Kluwer Academic, Dordrecht, 1994), .
150. P.F. Barbara, W.H. Knox, G.A. Mourou, and A.H. Zewail (eds.), *Ultrafast Phenomena IX*, Vol. 60 *Springer Series in Chemical Physics*, (Springer, Berlin, Heidelberg, 1995).
151. D.E. Spence, P.N. Kean, and W. Sibett, '60-fsec Pulse Generation from a Self-Modelocked Ti:Sapphire Laser', Opt. Lett. **16**, 42 (1991).
152. U. Keller, G.W. 'tHooft, W.H. Knox, and J.E. Cunningham, 'Femtosecond Pulses from a Continuously Self-Starting Passively Mode-Locked Ti:Sapphire Laser', Opt. Lett. **16**, 1022 (1991).

153. C. Spielmann, M. Lenzer, F. Krausz, R. Szipócs, and K. Ferencz, 'Chirped Dielectric Mirrors Improve Ti:Sapphire Lasers', Focus World **12**, 55 (1995).
154. E.B. Treacy, 'Compression of Picosecond Light Pulses', Phys. Lett. A **28**, 34 (1968).
155. R.L. Fork, O.E. Martinez, and J.P. Gordon, 'Negative Dispersion Using Pairs of Prisms', Opt. Lett. **9**, 150 (1984).
156. J.A. Giordmaine, M.A. Duguay, and J.W. Hansen, 'Compression of Optical Pulses', IEEE J. Quant. Electr. **QE-4**, 252 (1968).
157. J.A. Valdmanis, R.L. Fork, and J.P. Gordon, 'Generation of Optical Pulses as Short as 27 Femtoseconds Directly from Laser Balancing Self-Phase Modulation, Group-Velocity Dispersion, Saturable Absorption, and Saturable Gain', Opt. Lett. **10**, 131 (1985).
158. W. Rudolph and B. Wilhelmi, *Light Pulse Compression*, Vol. 3 *Laser Science and Technology* (Harwood Academic, Chur, 1989).
159. J.D. Kafka, M.L. Watts, and J.-W. Pieterse, 'Picosecond and Femtosecond Pulse Generation in a Regeneratively Mode-Locked Ti:Sapphire Laser', IEEE J. Quant. Electr. **QE-28**, 2151 (1992).
160. J. Goodberlet, J. Wang, J.G. Fujimoto, and P.A. Schulz, 'Femtosecond Passively Mode-Locked Ti:Al_2O_3 Laser with a Nonlinear External Cavity', Opt. Lett. **14**, 1125 (1989).
161. M.T. Asaki, C.-P. Huang, D. Garvey, J. Zhou, H.C. Kapteyn, and M.M. Murnane, 'Generation of 11-fs Pulses from a Self-Mode-Locked Ti:Sapphire Laser', Opt. Lett. **18**, 977 (1993).
162. A. Stingl, M. Lenzner, C. Spielmann, F. Krausz, and R. Szipócs, 'Sub-10-fs Mirror Dispersion-Controlled Ti:Sapphire Laser', Opt. Lett. **20**, 602 (1995).
163. R.L. Fork, B.I. Greene, and C.V. Shank, 'Generation of Optical Pulses Shorter than 0.1 psec by Colliding Pulse Mode Locking', Appl. Phys. Lett. **38**, 671 (1981).
164. P. Simon, S. Smatmári, and F.P. Schäfer, 'Generation of 30-fs Pulses Tunable over the Visible Spectrum', Opt. Lett. **16**, 1569 (1991).
165. B. Wilhelmi, 'Propagation of Femtosecond Light Pulses Through Dye Amplifiers' in *Dye Lasers: 25 Years*, Vol. 70 *Topics in Applied Physics*, M. Stuke (ed.) (Springer, Berlin, Heidelberg, 1992), p 111.
166. F.K. Tittel, T. Hofmann, T.E. Sharp, P.J. Wisoff, W.L. Wilson, and G. Szabó, 'Blue-Green Dye Laser Seeded Operation of a Terawatt Excimer Amplifier' in *Dye Lasers: 25 Years*, Vol. 70 *Topics in Applied Physics*, M. Stuke (ed.) (Springer, Berlin, Heidelberg, 1992), p 141.
167. *The Supercontinuum Laser*, R.R. Alfano (ed.) (Springer, NewYork, 1989).
168. V. Petrov, D. Georgiev, and U. Stamm, 'Improved Mode Locking of a Femtosecond Titanium-Doped Sapphire Laser by Intracavity Second Harmonic Generation', Appl. Phys. Lett. **60**, 1550 (1992).
169. R.J. Ellingson and C.L. Tang, 'High-Repetion-Rate Femtosecond Pulse Generation in the Blue', Opt. Lett. **17**, 343 (1992).
170. S. Backus, M.T. Asaki, C. Shi, H.C. Kapteyn, and M.M. Murnane, 'Intracavity Frequency Doubling in a Ti:Sapphire Laser: Generation of 14-fs Pulses at 416 nm', Opt. Lett. **19**, 399 (1994).
171. W.S. Pelouch, P.E. Powers, and C.L. Tang, 'Ti:Sapphire-Pumped, High Repetion-Rate Femtosecond Optical Parametric Oscillator', Opt. Lett. **17**, 1070 (1992).
172. J. Watson, T. Lépine, P. Georges, and A. Brun, 'Tunable Femtosecond Pulse in the Visible' in *Ultrafast Phenomena IX*, Vol. 60 *Springer Series in Chemical Physics*, P.F. Barbara, W.H. Knox, G.A. Mourou, and A.H. Zewail (eds.) (Springer, Berlin, Heidelberg, 1995), p 178.

173. F. Seifert, V. Petrov, and M. Woerner, 'Solid-State Laser System for the Generation of Midinfrared Femtosecond Pulses Tunable from 3.3 to 10 μm', Opt. Lett. **19**, 2009 (1994).
174. J. Squier, F. Salin, G. Mourou, and D. Harter, '100-fs Pulse Generation and Amplification in $Ti:Al_2O_3$', Opt. Lett. **16**, 324 (1991).
175. F. Salin, J. Squier, G. Mourou, and G. Vaillancourt, 'Multikilohertz $Ti:Al_2O_3$ Amplifier for High-Power Femtosecond Pulses', Opt. Lett. **16**, 1964 (1991).
176. J. Squier, G. Korn, G. Mourou, G. Vaillancourt, and M. Bouvier, 'Amplification of Femtosecond Pulses at 10-kHz Repetition Rates in $Ti:Al_2O_3$', Opt. Lett. **18**, 625 (1993).
177. T.B. Norris, 'Femtosecond Pulse Amplification at 250 kHz with a Ti:Sapphire Regenerative Amplifier and Application to Continuum Generation', Opt. Lett. **17**, 1009 (1992).
178. E. Schreiber, *Femtosecond Real-Time Spectroscopy of Small Molecules and Clusters*, Habilitation thesis, Freie Universität Berlin, Berlin-Dahlem, 1996.
179. P.W. Smith, 'Mode-Locking of Lasers', Proc. IEEE **58**, 1342 (1970).
180. S.E. Harris and O.P. McDuff, 'Theory of FM Laser Oscillation', IEEE Quant. Electr. **1**, 245 (1965).
181. D.J. Kuizenga and A.E. Siegman, 'FM and AM Mode Locking of the Homogeneous Laser – Part I: Theory', IEEE Quant. Electr. **6**, 694 (1970).
182. G.H.C. New, 'The Generation of Ultrashort Laser Pulses', Rep. Prog. Phys. **46**, 877 (1983).
183. E.P. Ippen and C.V. Shank, 'Techniques and Measurement' in *Ultrashort Light Pulses, Picosecond Techniques and Applications*, Vol. 18 *Topics in Applied Physics*, S.L. Shapiro (ed.) (Springer, Berlin, Heidelberg, 1977), p 83.
184. C.V. Shank, 'Generation of Ultrashort Optical Pulses' in *Ultrashort Laser Pulses and Applications*, Vol. 60 *Topics in Applied Physics*, W. Kaiser (ed.) (Springer, Berlin, Heidelberg, 1988), p 5.
185. J.-C. Diels and W. Rudolph, *Ultrashort Laser Pulse Phenomena, Optics and Photonics Series* (Academic Press, San Diego, 1996).
186. P.M.W. French, 'The Generation of Ultrashort Laser Pulses', Rep. Prog. Phys. **58**, 169 (1996).
187. J.B. Atkinson, J. Becker, and W. Demtröder, 'Cavity Length Detuning Characteristics of the Synchronously Mode-Locked CW Dye Laser', IEEE Quant. Electr. **15**, 912 (1979).
188. D.J. Bradley, 'Methods of Generation' in *Ultrashort Light Pulses, Picosecond Techniques and Applications*, Vol. 18 *Topics in Applied Physics*, S.L. Shapiro (ed.) (Springer, Berlin, Heidelberg, 1977), p 17.
189. A.E. Siegman, *Lasers* (Oxford University Press, Oxford, 1986).
190. H. Kogelnik, E.P. Ippen, A. Dienes, and C.V. Shank, 'Astigmatically Compensated Cavities for CW Dye Lasers', IEEE Quant. Electr. **8**, 373 (1972).
191. P.F. Moulton, 'Ti-Doped Sapphire: a Tunable Solid-State Laser', Opt. News **8**, 9 (1982).
192. P.F. Moulton, 'Spectrocopic and Laser Characteristics of $Ti:Al_2O_3$', J. Opt. Soc. Am. B **3**, 125 (1986).
193. P. Albers, E. Stark, and G. Huber, 'Continuous-Wave Laser Operation and Quantum Efficiency of Titanium-Doped Sapphire', J. Opt. Soc. Am. B **3**, 134 (1986).
194. Y.R. Shen, 'Self-Focusing: Experimental', Prog. Quant. Electr. **4**, 1 (1975).
195. J.H. Marburger, 'Self-Focusing: Theory', Prog. Quant. Electr. **4**, 35 (1975).
196. P. Tournois, 'Negative Group Delay Times in Frustrated Gires–Tournois and Fabry–Perot Interferometers', IEEE Quant. Electr. **33**, 519 (1997).

197. F. Gires and O. Tournois, 'Interféromètre utilisable pour la compression d'impulsions lumineuses modulées en fréquence', C. R. Acad. Sci. Paris **258**, 6112 (1964).
198. J. Kuhl and J. Heppner, 'Compression of Femtosecond Optical Pulses with Dielectric Multilayer Interferometers', IEEE J. Quant. Electr. **22**, 182 (1986).
199. M.A. Duguay and J.W. Hansen, 'Compression of Pulses from a Mode Locked He–Ne Laser', Appl. Phys. Lett. **14**, 14 (1969).
200. H. A. Haus, J.G. Fujimoto, and E.P. Ippen, 'Analytic Theory of Additive Pulse and Kerr Lens Mode Locking', IEEE Quant. Electr. **28**, 2086 (1992).
201. F. Krausz, M.E. Fermann, T. Brabec, P.F. Curley, M. Hofer, M.H. Ober, C. Spielmann, E. Winter, and A.J. Schmidt, 'Femtosecond Solid-State Lasers', IEEE Quant. Electr. **28**, 2097 (1992).
202. U. Keller, W.H. Knox, and G.W. 'tHooft, 'Ultrafast Solid-State Mode-Locked Lasers Using Resonant Nonlinearities', IEEE Quant. Electr. **28**, 2133 (1992).
203. H. Ruppe, *Wellenpaketdynamik in dissoziativen Systemen*, Master's thesis, Freie Universität Berlin, Berlin-Dahlem, 1995.
204. J.A. Giordmaine and R.C. Miller, 'Tunable Coherent Parametric Oscillations in $LiNbO_3$ at Optical Frequencies', Phys. Rev. Lett. **14**, 973 (1965).
205. A. Yariv and J.E. Pearson, 'Parametric Process', Progr. Quant. Opt. **1**, 1 (1969).
206. *Handbook of Nonlinear Optics*, Vol. 52 *Optical Engineering*, R.L. Sutherland (ed.) (Marcel Decker Inc., New York, 1996), .
207. G.M. Gale, M. Cavallari, T.J. Driscoll, and F. Hache, 'Sub-20-fs Tunable Pulses in the Visible from an 82-MHz Optical Parametric Oscillator', Opt. Lett. **20**, 1562 (1995).
208. C. Chen, B. Wu, A. Jiang, G. You, R. Li, and S. Lin, 'New Nonlinear Optical Crystal: LiB_3O_5', J. Opt. Soc. Am. B **6**, 616 (1989).
209. C.L. Tang, W.R. Rosenberg, T. Ukachi, R.J. Lane, and L.K. Cheng, 'Optical Parametric Oscillators', Proc. IEEE **80**, 365 (1992).
210. C.T. Chen, *Development of New Nonlinear Optical Crystals in the Borate Series*, Vol. 15 *Laser Science and Technology: an International Handbook* (Harwood Academic, Chur, Switzerland, 1993).
211. T. Schröder, K.-J. Boller, A. Fix, and R. Wallenstein, 'Spectral Properties and Numerical Modelling of a Critically Phase-Matched Nanosecond LiB_3O_5 Optical Parametric Oscillator', Appl. Phys. B **58**, 425 (1994).
212. R.L. Byer, *Nonlinear Optics,* P.G. Harper and B.S. Wherret (eds.) (Academic Press, London, 1977).
213. A. Yariv (ed.), *Optical Electronics*, (Saunders College Publishing, Philadelphia, 1988).
214. N. Bloembergen, *Nonlinear Optics* (World Scientific, Singapore, 1996).
215. X.M. Zhao and D.J. McGraw, 'Parametric Mode Locking', IEEE Quant. Electr. **28**, 930 (1992).
216. M.J. McCarthy and D.C. Hanna, 'Continuous-Wave Mode-Locked Singly Resonant Optical Parametric Oscillator Synchronously Pumped by a Laser-Diode-Pumped Nd:YLF Laser', Opt. Lett. **17**, 402 (1992).
217. M. Ebrahimzadeh, G.P.A. Malcolm, and A.I. Ferguson, 'Continuous-Wave Mode-Locked Optical Parametric Oscillator Synchronously Pumped by a Diode-Laser-Pumped Solid-State Laser', Opt. Lett. **17**, 183 (1992).
218. M. Ebrahimzadeh, S. French, W. Sibett, and A. Müller, 'Picosecond Ti:Sapphire Optical Parametric Oscillator Based on LiB_3O_5', Opt. Lett. **20**, 166 (1995).

219. E.S. Wachmann, D.C. Edelstein, and C.L. Tang, 'Continuous-Wave Mode-Locked and Dispersion-Compensated Femtosecond Optical Parametric Oscillator', Opt. Lett. **15**, 136 (1990).
220. A. Fix, T. Schröder, and R. Wallenstein, 'New Sources of Powerful Tunable Laser Radiation in the Ultraviolet, Visible and Near Infrared', Laser und Optoelektronik **23**, 1991 (1991).
221. S. French, M. Ebrahimzadeh, and A. Miller, 'High-Power, High-Repetition-Rate Picosecond Optical Parametric Oscillator Tunable in the Visible', Opt. Lett. **21**, 976 (1996).
222. K. Kobe, *Ultrakurzzeit-Spektroskopie der intramolekularen und dissoziativen Dynamik von Alkalimetallclustern*, PhD thesis, Freie Universität Berlin, Berlin-Dahlem, 1993.
223. E. Schreiber, 'Femtosecond Real-Time Spectroscopy of Small Molecules and Clusters' in *Proceedings of the International Conference on LASERS '94*, V.J. Corcoran and T. Goldman (eds.) (Society for Optical and Quantum Electronics, McLean (VA), 1995), p 490.
224. D. von der Linde, 'Characterization of the Noise in Continuously Mode-Locked Lasers', Appl. Phys. **39**, 201 (1986).
225. K.L. Sala, G.A. Kenney-Wallace, and G.E. Hall, 'Cw Autocorrelation Measurements of Picosecond Laser Pulses', IEEE J. Quant. Electr. **16**, 990 (1980).
226. R. Trebino and D.J. Kane, 'Using Phase Retrieval to Measure the Intensity and Phase of Ultrashort Pulses: Frequency-Resolved Optical Gating', J. Opt. Soc. Am. A **10**, 1101 (1993).
227. D.J. Kane and R. Trebino, 'Characterization of Arbitary Femtosecond Pulses Using Frequency-Resolved Optical Gating', IEEE Quant. Electr. **29**, 571 (1993).
228. K.W. DeLong, D.N. Fittinghoff, and R. Trebino, 'Practical Issues in Ultrashort-Laser-Pulse Measurement Using Frequency-Resolved Optical Gating', IEEE Quant. Electr. **32**, 1253 (1996).
229. M.A. Krumbügel, C.L.Ladera, K.W. DeLong, D.N. Fittinghoff, J.N. Sweetser, and R. Trebino, 'Direct Ultrashort-Pulse Intensity and Phase Retrieval by Frequency-Resolved Optical-Gating and a Computational Neural Network', Opt. Lett. **21**, 143 (1996).
230. B. Kohler, V.V. Yakovlev, K.R. Wilson, K.W. DeLong, and R. Trebino, 'Phase and Intensity Characterization of Femtosecond Pulses from a Chirped-Pulse Amplifier by Frequency-Resolved Optical-Gating', Opt. Lett. **20**, 483 (1995).
231. G. Taft, A. Rundquist, M.M. Murnane, H.C. Kapteyn, K.W. DeLong, R. Trebino, and P. Christov I, 'Ultrashort Optical Waveform Measurements Using Frequency-Resolved Optical Gating', Opt. Lett. **20**, 743 (1995).
232. K.W. DeLong, R. Trebino, and W.E. White, 'Simultaneous Recovery of Two Ultrashort Laser Pulses from a Single Spectrogram', J. Opt. Soc. Am. B **12**, 2463 (1995).
233. K.W. DeLong, C.L. Ladera, R. Trebino, B. Kohler, and K.R. Wilson, 'Ultrashort-Pulse Measurement Using Noninstantaneous Nonlinearities: Raman Effects in Frequency-Resolved Optical Gating', Opt. Lett. **20**, 486 (1995).
234. D.N. Fittinghoff, K.W. DeLong, and R. Trebino, 'Noise Sensitivity in Frequency-Resolved Optical-Gating Measurements of Ultrashort Pulses', J. Opt. Soc. Am. B **12**, 1955 (1995).
235. K.W. DeLong, R. Trebino, J. Hunter, and W.E. White, 'Frequency-Resolved Optical Gating with the Use of Second-Harmonic Generation', J. Opt. Soc. Am. B **11**, 2206 (1994).

236. K.W. DeLong and R. Trebino, 'Improved Ultrashort Pulse-Retrieval Algorithm for Frequency-Resolved Optical Gating', J. Opt. Soc. Am. A **11**, 2429 (1994).
237. K.W. DeLong, R. Trebino, and D.J. Kane, 'Comparison of Ultrashort-Pulse Frequency-Resolved Optical Gating Traces for Three Common Beam Geometries', J. Opt. Soc. Am. B **11**, 2206 (1994).
238. H. Haberland, 'Experimental Methods' in *Clusters of Atoms and Molecules II*, Vol. 56 *Springer Series in Chemical Physics*, H. Haberland (ed.) (Springer, Berlin, Heidelberg, 1994), p 207.
239. D.R. Miller, 'Free Jet Sources' in *Atomic and Molecular Beam Methods*, G. Scoles, D. Bassi, U. Buck, and D. Lainé (eds.) (Oxford University Press, New York, 1988), pp. 14–53.
240. O. F. Hagena, 'Nucleation and Growth of Clusters in Expanding Nozzle Flows', Surf. Sci. **106**, 101 (1981).
241. O. F. Hagena, 'Condensation in Free Jets: Comparison of Rare Gases and Metals', Z. Phys. D **4**, 291 (1987).
242. W. Demtröder and H.-J. Foth, 'Molekülspektroskopie in kalten Düsenstrahlen', Phys. Bl. **43**, 7 (1987).
243. M.A.D. Fluendy and K.P. Lawley, *Chemical Applications of Molecular Beam Scattering* (Chapman & Hall, London, 1973).
244. J.B. Anderson, R.P. Andres, and J.B. Fenn, 'Supersonic Nozzle Beams', Adv. Chem. Phys. **10**, 275 (1966).
245. J.B. Anderson and J.B. Fenn, 'Velocity Distributions in Molecular Beams from Nozzle Sources', Phys. Fluids **8**, 780 (1965).
246. S.B. Ryali and J.B. Fenn, 'Clustering in Free Jets – Aggregation by Dispersion', Ber. Bunsenges. Phys. Chem. **88**, 245 (1984).
247. J. Fricke, 'Kondensation in Düsenstrahlen', Physik in unserer Zeit **4**, 21 (1973).
248. G. Delacrétaz and L. Wöste, 'Vibrational Resolved Spectroscopy of Sodium Clusters', Surf. Sci. **156**, 770 (1985).
249. M. Ulbricht, *Hochauflösende Laserspektroskopie an kleinen Alkaliclustern*, Master's thesis, Freie Universität Berlin, Berlin-Dahlem, 1991.
250. T. Bocher, *Zeke-Photoelektronenspektrometer und Langmuir-Taylor-Detektor*, Master's thesis, Freie Universität Berlin, Berlin-Dahlem, 1992.
251. W. Paul, H.P. Reinhard, and U. von Zahn, 'Das elektrische Massenfilter als Massenspektrometer und Isotopentrenner', Z. Phys. **152**, 143 (1958).
252. D. Bassi, 'Ionization Detectors II: Mass Selection and Ion Detection' in *Atomic and Molecular Beam Methods*, G. Scoles, D. Bassi, U. Buck, and D. Lainé (eds.) (Oxford University Press, New York, 1988), p 180.
253. N.F. Ramsey, *Molecular Beams* (Oxford University Press, Oxford, 1985).
254. J.B. Taylor, 'Eine Methode zur direkten Messung der Intensitätsverteilung in Molekularstrahlen', Zeitschrift für Physik **57**, 242 (1929).
255. J.B. Taylor, 'The Reflection of Beams of the Alkali Metals from Crystals', Phys. Rev. **35**, 375 (1930).
256. I. Langmuir and K.H. Kingdon, 'Thermionic Effects Caused by Alkali Vapors in Vacuum Tubes', Science **57**, 58 (1923).
257. I. Langmuir and K.H. Kingdon, 'Thermionic Phenomena due to Alkali Vapors. Part II: Theoretical', Phys. Rev. **21**, 381 (1923).
258. K.H. Kingdon and I. Langmuir, 'Thermionic Phenomena due to Alkali Vapors. Part I: Experimental', Phys. Rev. **21**, 380 (1923).
259. T.J. Killian, 'Thermionic Phenomea Caused by Vapors of Rubidium and Potassium', Phys. Rev. **27**, 578 (1926).

260. H. Ruppe, S. Rutz, E. Schreiber, and L. Wöste, 'Femtosecond Wave Packet Propagation Dynamics in the Dissociative K_3 Molecule', Chem. Phys. Lett. **257**, 356 (1996).
261. S. Rutz, *Zeitaufgelöste Spektroskopie am B-Zustand des Na_3*, Master's thesis, Freie Universität Berlin, Berlin-Dahlem, 1992.
262. S. Rutz, *Femtosekundenspektroskopie zur Wellenpaketdynamik in Alkalidimeren und -trimeren*, Ph.D. thesis, Freie Universität Berlin, Berlin-Dahlem, 1996.
263. E. Schreiber, 'Laser-Femtochemistry of Small Clusters' in *Theory of Atomic and Molecular Clusters*, J. Jellinek (ed.) (Springer, Berlin, Heidelberg, in press).
264. D. Strickland and G. Mourou, 'Compression of Amplified Chirped Optical Pulses', Opt. Commun. **56**, 219 (1985).
265. P. Maine, D. Strickland, P. Bado, M. Pessot, and G. Mourou, 'Generation of Ultrahigh Peak Power Pulses by Chirped Pulse Generation', IEEE J. Quant. Electr. **24**, 398 (1988).
266. Y.-H. Chuang, L. Zheng, and D.D. Meyerhofer, 'Propagation of Light Pulses in a Chirped-Pulse Amplification Laser', IEEE Quant. Electr. **29**, 270 (1993).
267. F. Salin, J. Squier, and G. Mourou, 'Large Temporal Stretching of Ultrashort Pulses', Appl. Opt. **31**, 1225 (1992).
268. J. Squier, S. Coe, K. Clay, G. Mourou, and D. Harter, 'An Alexandrite Pumped Nd:Glass Regenerative Amplifier for Chirped Pulse Amplification', Opt. Comm. **92**, 73 (1992).
269. G. Vaillancourt, J.S. Coe, P. Bado, and G.A. Mourou, 'Operation of a 1 kHz Pulse-Pumped Ti:Sapphire Regenerative Amplifier', Opt. Lett. **15**, 317 (1990).
270. *Manual Tw-OPA/OPG 'TOPAS'* (distributed by Quantronix, New York, 1996).
271. S. Wolf, *Zeitaufgelöste Spektroskopie an Silberclustern*, Ph.D. thesis, Freie Universität Berlin, Berlin-Dahlem, 1997.
272. R.J. Colbton, M.M. Ross, and D.A. Kidwell, 'Secondary Ion Mass Spectrometry: Polyatomic and Molecular Ion Emission', Nucl. Instr. Meth. Phys. Res. B **13**, 259 (1986).
273. B. Garrison, N. Winograd, and D. Harrison Jr., 'Formation of Small Metal Clusters by Ion Bombardment of Single Crystal Surfaces', J. Chem. Phys. **69**, 1440 (1978).
274. P. Sigmund, 'Theory of Sputtering. I. Sputtering Yield of Amorphous and Polycrystalline Targets', Phys. Rev. **184**, 383 (1969).
275. R. Keller, F. Nöhmeier, P. Spädtke, and M.H. Schönenberg, 'CORDIS – an Improved High-Current Ion Source for Gases', Vacuum **34**, *Nos. 1-2*, 31 (1984).
276. R. Keller, 'High-Current Gaseous Ion Sources' in *The Physics and Technology of Ion Sources*, J.G. Brown (ed.) (Wiley, New York, 1989), p 151.
277. V.E. Krohn, 'Emission of Negative Ions from Metal Surfaces Bombarded by Positive Cesium Ions', J. Appl. Phys. **33**, 3523 (1962).
278. G. Hortig, P. Mokler, and M. Müller, 'Eine Quelle für starke Ströme negativer schwerer Ionen', Z. Phys. **210**, 312 (1968).
279. L. Hanley, S.A. Ruatta, and S.L. Anderson, 'Collision-Induced Dissociation of Aluminum Cluster Ions: Fragmentation Patterns, Bond Energies, and Structures for Al_2^+–Al_7^+', J. Chem. Phys. **87**, 260 (1987).
280. G.G. Dolnikowski, M.J. Kristo, C.G. Enke, and J.T. Watson, 'Ion-Trapping Technique for Ion/Molecule Reaction Studies in the Center Quadrupole of a Triple Quadrupole Mass Spectrometer', Int. J. Mass Spectr. Ion Proc. **82**, 1 (1988).
281. P.Y. Cheng and M.A. Duncan, 'Vibronic Spectroscopy and Dynamics in the Jet-Cooled Silver Trimer', Chem. Phys. Lett. **152**, 341 (1988).

282. A.M. Ellis, E.S.J. Robles, and T.A. Miller, 'Dispersed Fluorescence Spectroscopic Study of the Ground Electronic State of Silver Trimer', Chem. Phys. Lett. **201**, 132 (1993).
283. J. Jortner and R.S. Berry, 'Radiationless Transitions and Molecular Quantum Beats', J. Chem. Phys. **48**, 2757 (1968).
284. E.J. Heller, 'The Semiclassical Way to Molecular Spectroscopy', Acc. Chem. Res. **14**, 368 (1981).
285. M.D. Feit, J.A. Fleck Jr., and A. Steiger, 'Solution of the Schrödinger Equation by a Spectral Method', J. Comp. Phys. **47**, 412 (1982).
286. M.D. Feit and J.A. Fleck Jr., 'Solution of the Schrödinger Equation by a Spectral Method II. Vibrational Energy Levels of Triatomic Molecules', J. Chem. Phys. **78**, 301 (1983).
287. D. Kosloff and R. Kosloff, 'A Fourier Method Solution for the Time Dependent Schrödinger Equation as Tool in Molecular Dynamics', J. Comput. Phys. **52**, 35 (1983).
288. R. Kosloff and D. Kosloff, 'A Fourier Method Solution for the Time Dependent Schrödinger Equation: A Study of the Reaction $H^{+}+H_2$, $D^{+}+HD$, and $D^{+}+H_2$', J. Chem. Phys. **79**, 1823 (1983).
289. G.K. Paramonov and V.A. Savva, 'Resonance Effects in Molecule Vibrational Excitation by Picosecond Laser Pulses', Phys. Lett. A **97**, 340 (1983).
290. T. Joseph and J. Manz, 'Mode Selective Dissociation of Vibronically Excited ABA Molecular Resonances Stimulated by a Picosecond Infrared Laser Pulse', Molec. Phys. **58**, 1149 (1986).
291. V. Engel and H. Metiu, 'CH_3ONO Predissociation by Ultrashort Laser Pulses: Population Transients and Product State Distribution', J. Chem. Phys. **92**, 2317 (1990).
292. H. Metiu and V. Engel, 'A Theoretical Study of I_2 Vibrational Motion after Excitation with an Ultrashort Pulse', J. Chem. Phys. **93**, 5693 (1990).
293. J. Manz, B. Reischl, T. Schröder, and F. Seyl, 'On the Laser-Femtochemistry Approach to Coherent Molecular Vibrations. Model Simulations for $Ni[C_2D_4]\rightarrow Ni+C_2D_4$', Chem. Phys. Lett. **198**, 483 (1992).
294. B. Hartke, R. Kosloff, and R. Ruhman, 'Large Amplitude Ground State Vibrational Coherence Induced by Impulsive Absorption in CsI. A Computer Simulation', Chem. Phys. Lett. **158**, 238 (1989).
295. R. de Vivie-Riedle, *Theoretisch-chemische Untersuchungen zur Spektroskopie und Dynamik kleiner Moleküle und Cluster*, Habilitation thesis, Freie Universität Berlin, Berlin-Dahlem, 1997.
296. H. Tal-Ezer and R. Kosloff, 'An Accurate and Efficient Scheme for Propagating the Time Dependent Schrödinger Equation', J. Chem. Phys. **81**, 3967 (1984).
297. D.T. Colbert and W.H. Miller, 'A Novel Discrete Variable Representation for Quantum Mechanical Reactive Scattering via the S-Matrix Kohn Method', J. Chem. Phys. **96**, 1982 (1992).
298. W.H. Press, B.P. Flannery, S.A. Teukolsky, and W.T. Vetterling, *Numerical Recipes* (Cambridge University Press, Cambridge, 1986).
299. R. Kosloff, 'Time-Dependent Quantum-Mechanical Methods for Molecular Dynamics', J. Phys. Chem. **92**, 2087 (1988).
300. C. Leforestier, R. Bisseling, C. Cerjan, M.D. Feit, R. Friesner, A. Gulberg, A. Hammerich, G. Jolicard, W. Karrlein, H.-D. Meyer, N. Lipken, O. Roncero, and R. Kosloff, 'A Comparison of Different Propagation Schemes for the Time Dependent Schrödinger Equation', J. Comput. Phys. **94**, 59 (1991).
301. R. Kosloff, 'Propagation Methods for Quantum Molecular Dynamics', Ann. Rev. Phys. Chem. **45**, 145 (1994).

302. M. Seel and W. Domcke, 'Femtosecond Time-Resolved Ionization Spectroscopy of Ultrafast Internal Conversion Dynamics in Polyatomic Molecules: Theory and Computational Studies', J. Chem. Phys. **95**, 7806 (1991).
303. B.W. Shore, 'Coherence in the Quasi-Continuum Model', Chem. Phys. Lett. **99**, 240 (1983).
304. R.S. Burkey and C.D. Cantrell, 'Discretization in the Quasi-Continuum', J. Opt. Soc. Am. B **1**, 169 (1984).
305. R.S. Burkey and C.D. Cantrell, 'Multichannel Excitation of the Quasi-Continuum', J. Opt. Soc. Am. B **2**, 451 (1985).
306. E. Schreiber, H. Kühling, K. Kobe, S. Rutz, and L. Wöste, 'Time-Resolved TPI-Spectroscopy of the B- and D-State of Na_3-Clusters', Ber. Bunsenges. Phys. Chem. **96**, 1301 (1992).
307. A. Ruff, S. Rutz, E. Schreiber, and L. Wöste, 'Ultrafast Photodissociation of $K_{n=3...9}$ Clusters', Z. Phys. D **37**, 175 (1996).
308. J.A. Nelder and R. Mead, 'A Simplex Method for Function Minimization', Comput. J. **7**, 308 (1965).
309. M. Warken and V. Bonačić-Koutecký, 'Quantum Mechanical Treatment of Stationary and Dynamical Properties of Bound Vibrational Systems. Application to the Relaxation Dynamics of Ag_5 after an Electron Photodetachment', Chem. Phys. Lett. **272**, 284 (1997).
310. D. Papousek and M.R. Aliev, *Molecular Vibrational Rotational Spectra* (Elsevier, Amsterdam, 1982).
311. A. Nauts and R.E. Wyatt, 'New Approach to Many-State Quantum Dynamics: The Recursive-Residue-Generation Method', Phys. Rev. Lett. **51**, 2238 (1983).
312. O. Rubner, C. Meier, and V. Engel, 'The Calculation of Time-Resolved Negative-Ion-to-Neutral-to-Positive-Ion Spectra with an Application to Iron-Carbonyl', J. Chem. Phys. **107**, 1066 (1997).
313. V. Engel, 'Femtosecond Pump/Probe Experiments and Ionization: the Time Dependence of the Total Ion Signal', Chem. Phys. Lett. **178**, 130 (1990).
314. J. Heufelder, *Langzeitdynamik der Femtosekunden-Wellenpaketpropagation im A-Zustand von NaK*, Master's thesis, Freie Universität Berlin, Berlin-Dahlem, 1996.
315. T.S. Rose, M. Rosker, and A.H. Zewail, 'Femtosecond Real-Time Observation of Wave Packet Oscillations (Resonances) in Dissociation Reactions', J. Chem. Phys. **88**, 6672 (1988).
316. S.I. Ionov, G.A. Brucker, C. Jaques, L. Valachovic, and C. Wittig, 'Subpicosecond Resolution Studies of the $H+CO_2 \rightarrow CO+OH$ Reaction Photoinitiated in CO_2–HI Complexes', J. Phys. Chem. **99**, 6553 (1984).
317. J.H. Glowina, J. Misewich, and P.P. Sorokin, 'Utilization of UV and IR Supercontinua in Gas-Phase Subpicosecond Kinetic Spectroscopy' in *The Supercontinuum Laser*, R.R. Alfano (ed.) (Springer, NewYork, 1989).
318. Y. Chen, L. Hunziker, P. Ludowise, and M. Morgen, 'Femtosecond Transient Stimulated Emission Pumping Studies of Ozone Visible Photodissociation', J. Chem. Phys. **97**, 2149 (1992).
319. J.B. Atkinson, J. Becker, and W. Demtröder, 'Experimental Observation of the $a^3\Pi_u$ State of Na_2', Chem. Phys. Lett. **87**, 92 (1982).
320. C. Effantin, O. Babaky, K. Hussein, J. d'Incan, and R.F. Barrow, 'Interactions between the $A^1\Sigma_u^+$ and $b^3\Pi_u$ States of Na_2', J. Phys. B **18**, 4077 (1985).
321. A.J. Ross, P. Crozet, C. Effantin, J. d'Incan, and R.F. Barrow, 'Interactions between the $A(1)^1\Sigma_u^+$ and $b(1)\,^3\Pi_u$ States of K_2', J. Phys. B **20**, 6225 (1987).
322. J. Heinze, U. Schühle, F. Engelke, and C.D. Cadwell, 'Doppler-Free Polarization Spectroscopy of the $B^1\Pi_u - X^1\Sigma_g^+$ Band System of K_2', J. Chem. Phys. **87**, 45 (1987).

323. A.M. Lyyra, W.T. Luh, L. Li, H. Wang, and W.C. Stwalley, 'The $A^1\Sigma_u^+$ State of the Potassium Dimer', J. Chem. Phys. **92**, 43 (1990).
324. G. Jong, L. Li, T.-J. Whang, and W.C. Stwalley, 'Cw All-Optical Triple Resonance Spectroscopy of K_2: Deperturbation Analysis of the $A^1\Sigma_u^+$ ($v \leq 12$) and $b^3\Pi_u$ ($13 \leq v \leq 24$) States', J. Mol. Spectrosc. **155**, 115 (1992).
325. W. Meyer, Universität Kaiserslautern, private communication.
326. M. Broyer, J. Chevaleyre, G. Delacrétaz, S. Martin, and L. Wöste, 'K_2 Rydberg State Analysis by Two- and Three-Photon Ionization', Chem. Phys. Lett. **99**, 206 (1983).
327. I. Fischer, M.J.J. Vrakking, D.M. Villeneuve, and A. Stolow, 'Femtosecond Time-Resolved Zero Kinetic Energy Photoelectron and Photoionization Spectroscopy Studies of I_2 Wave Packet Dynamics', Chem. Phys. **207**, 331 (1996).
328. R. de Vivie-Riedle, K. Kobe, J. Manz, W. Meyer, B. Reischl, S. Rutz, E. Schreiber, and L. Wöste, 'Femtosecond Study of Multiphoton Ionization Processes in K_2: from Pump-Probe to Control', J. Phys. Chem. **100**, 7789 (1996).
329. M. Rosker, T.S. Rose, and A.H. Zewail, 'Femtosecond Real-Time Dynamics of Photofragment-Trapping Resonances on Dissocative Potential Energy Surfaces', Chem. Phys. Lett. **146**, 175 (1988).
330. P. Kusch and M.M. Hessel, 'An Analysis of the $B^1\Pi_u - X^1\Sigma_g^+$ Band System of Na_2', J. Chem. Phys. **68**, 2591 (1978).
331. M.E. Kaminsky, 'New Spectroscopic Constants and RKR Potential for the $A^1\Sigma_u^+$ State of Na_2^*', J. Chem. Phys. **66**, 4951 (1977).
332. G. Gerber and R. Möller, 'Optical–Optical Double Resonance Spectroscopy of High Vibrational Levels of the Na_2 $A\,^1\Sigma_u^+$ State in a Molecular Beam', Chem. Phys. Lett. **113**, 546 (1984).
333. S. Rutz, S. Greschik, E. Schreiber, and L. Wöste, 'Femtosecond Wave Packet Propagation in Spin–Orbit Coupled Electronic States of the Na_2 Molecule', Chem. Phys. Lett. **257**, 365 (1996).
334. R.S. Mullikan, 'Role of Kinetic Energy in the Franck–Condon Principle', J. Chem. Phys. **55**, 309 (1971).
335. M. Machholm and A. Suzor-Weiner, 'Pulse length control of Na_2^+ photodissociation by intense femtosecond lasers', J. Chem. Phys. **105**, 971 (1996).
336. T.-J. Whang, H. Wang, A.M. Lyyra, L. Li, and W.C. Stwalley, 'Optical-Optical Double Resonance Spectroscopy of the Na_2 $2^1\Pi_g$ State', J. Mol. Spectrosc. **145**, 112 (1991).
337. S.A. Rice, 'Active Control of Selectivity of Product Formation in a Chemical Reaction: What's New?' in *Mode Selective Chemistry*, J. Jortner, R.D. Levine, and B. Pullman (eds.) (Kluwer Academic, Dordrecht, 1991), p 485.
338. E.D. Potter, J.L. Herek, S. Peterson, Q. Liu, and A.H. Zewail, 'Femtosecond Laser Control of Chemical Reaction', Nature **355**, 66 (1992).
339. S. Ruhman, A.G. Joly, and K.A. Nelson, 'Coherent Molecular Vibrational Motion Observed in the Time Domain Through Impulsive Stimulated Raman Scattering', IEEE J. Quant. Electr. **24**, 460 (1988).
340. W.T. Pollard, S.Y. Lee, and R.A. Mathies, 'Wave Packet Theory of Dynamic Absorption Spectra in Femtosecond Pump–Probe Experiments', J. Chem. Phys. **92**, 4012 (1990).
341. J. Chesnoy and A. Mokhtari, 'Resonant Impulsive-Stimulated Raman Scattering on Malachite Green', Phys. Rev. A **38**, 3566 (1988).
342. I.A. Wamsley, F.W. Wise, and C.L. Tang, 'On the Difference between Quantum Beats in Impulsive Stimulated Raman Scattering and Resonance Raman Scattering', Chem. Phys. Lett. **154**, 315 (1989).

343. Y.J. Yan and S. Mukamel, 'Femtosecond Pump–Probe Spectroscopy of Polyatomic Molecules in Condensed Phases', Phys. Rev. A **41**, 6485 (1990).
344. G. Stock and W. Domcke, 'Model Studies on the Time-Resolved Measurement of Excited-State Vibrational Dynamics and Vibronic Coupling', J. Opt. Soc. Am. B **7**, 1971 (1990).
345. S. Mukamel, 'Femtosecond Optical Spectroscopy: A Direct Look at Elementary Chemical Events', Annu. Rev. Phys. Chem. **41**, 647 (1990).
346. Y.J. Yan and S. Mukamel, 'Pulse Shaping and Coherent Raman Spectroscopy in Condensed Phases', J. Chem. Phys. **94**, 997 (1991).
347. D. Imre, J.L. Kinsey, A. Sinha, and J. Krenos, 'Chemical Dynamics Studied by Emission Spectroscopy of Dissociative Molecules', J. Phys. Chem. **88**, 3956 (1984).
348. A. Bartana, U. Banin, S. Ruhman, and R. Kosloff, 'Intensity Effects on Impulsive Excitation of Ground Surface Coherent Vibrational Motion. A 'V' Jump Simulation', Chem. Phys. Lett. **229**, 211 (1994).
349. R. Kosloff, A.D. Hammerich, and D.J. Tannor, 'Excitation without Demolition: Radiative Excitation of Ground Surface Vibration by Impulsive Stimulated Raman Scattering with Damage Control', Phys. Rev. Lett. **61**, 2172 (1992).
350. A. Bartana, U. Banin, S. Ruhman, and R. Kosloff, 'Laser Cooling of Molecular Internal Degrees of Freedom by a Series of Shaped Pulses', J. Chem. Phys. **99**, 196 (1993).
351. R. de Vivie-Riedle and B. Reischl, 'Quantum Calculations of Femtosecond Pump–Probe Spectroscopy in K_2 for Low Laser Field Intensities', Ber. Bunsenges. Phys. Chem. **99**, 485 (1995).
352. L. Allen and J.H. Eberly, *Optical Resonance and Two-Level Atoms* (Dover, New York, 1987).
353. L.-E. Berg, M. Beutter, and T. Hansson, 'Femtosecond Laser Spectroscopy on the Vibrational Wave Packet Dynamics of the $A^1\Sigma^+$ State of NaK', Chem. Phys. Lett. **253**, 327 (1996).
354. S. Rutz and E. Schreiber, 'Laser-Induced Ultrafast Vibrational Wave Packet Dynamics in the NaK $A^1\Sigma^+$ State' in *Proceedings of the International Conference on LASERS '96*, V.J. Corcoran and T. Goldman (eds.) (Society for Optical and Quantum Electronics, McLean, 1997), p 341.
355. I.Sh. Averbukh and N.F. Perel'man, 'The Dynamics of Wave Packets of Highly-Excited States of Atoms and Molecules', Sov. Phys. Usp. **34**, 572 (1991).
356. J. Parker and C.R. Stroud Jr., 'Coherence and Decay of Rydberg Wave Packets', Phys. Rev. Lett. **56**, 716 (1986).
357. A.J. Ross, R.M. Clements, and R.F. Barrow, 'The $A(2)^1\Sigma^+$ State of NaK', J. Mol. Spectrosc. **127**, 546 (1988).
358. A. Herrmann, E. Schuhmacher, and L. Wöste, 'Preparation and Photoionization Potentials of Molecules of Sodium, Potassium, and Mixed Atoms', J. Chem. Phys. **68**, 2327 (1978).
359. W.J. Stevens, D.D. Konowalow, and L.B. Ratcliff, 'Electronic Structure and Spectra of the Lowest Five $^1\Sigma^+$ and $^3\Sigma^+$ States, and Lowest Three $^1\Pi$, $^3\Pi$, $^1\Delta$, and $^3\Delta$ States of NaK', J. Chem. Phys. **80**, 1215 (1984).
360. A.J. Ross, C. Effantin, J. d'Incan, and R.F. Barrow, 'Long-Range Potentials for the $X^1\Sigma^+$ and $a^3\Sigma^+$ States of the NaK Molecule', Mol. Phys. **56**, 903 (1985).
361. A.J. Ross, C. Effantin, J. d'Incan, and R.F. Barrow, 'Laser-Induced Flourescence of NaK: the $b(1)^3\Pi$ State', J. Phys. B **19**, 1449 (1986).

362. A.J. Ross, *Etude par spectrométrie par transformation de Fourier des états électronique de basse énergie des molecules NaK et K_2*, PhD thesis, Université Claude Bernard-Lyon I, Lyon, 1987.
363. Ch. Meier, V. Engel, and J.S. Briggs, 'Long Time Wave Packet Behavior in a Curve-Crossing System: The Predissociation of NaI', J. Chem. Phys. **95**, 7337 (1991).
364. S.I. Vetchinkin, A.S. Vetchinkin, V.V. Eryomin, and I.M. Umanskii, 'Gaussian Wavepacket Dynamics in an Anharmonic System', Chem. Phys. Lett. **215**, 11 (1993).
365. S.I. Vetchinkin and V.V. Eryomin, 'The Structure of Wavepacket Fractional Revivals in a Morse-Like Anharmonic System', Chem. Phys. Lett. **222**, 394 (1994).
366. R.L. Martin and E.R. Davidson, 'Electronic Structure of the Sodium Trimer', Mol. Phys. **35**, 1713 (1978).
367. J.L. Martins, R. Car, and J. Buttet, 'Electronic Properties of Alkali Trimers', J. Chem. Phys. **78**, 5646 (1983).
368. T. Oka, 'Coherence and Decay of Rydberg Wave Packets', Phys. Rev. Lett. **45**, 351 (1980).
369. G. Delacrétaz, P. Fayet, J.P. Wolf, and L. Wöste, 'Optical and Dynamical Properties of Metal Clusters' in *Elemental and Molecular System*, Vol. 6 *Springer Series in Matrial Science*, G. Benedek, T.P. Martin, and G. Pacchioni (eds.) (Springer, Berlin, Heidelberg, 1988), p 64.
370. J.P. Wolf, G. Delacrétaz, and L. Wöste, 'First Observation of an Electronically Excited State of Li_3', Phys. Rev. Lett. **63**, 146 (1989).
371. W.H. Gerber, *Theorie des dynamischen Jahn-Teller Effektes in Li_3 und Untersuchung von Lithium-Molekularstrahlen*, Ph.D. thesis, Universität Bern, Bern, 1980.
372. A.J.C. Varandas and V.M.F. Morais, 'Semi-Empirical Valence Bond Potential Energy Surfaces for Homonuclear Alkali Trimers', Mol. Phys. **47**, 1241 (1982).
373. T.C. Thompson, G. Izmirlian Jr., S.J. Lemon, D.G. Truhlar, and C.A. Mead, 'Consistent Analytic Representation of the Two Lowest Potential Energy Surfaces for Li_3, Na_3, and K_3', J. Chem. Phys. **82**, 5597 (1985).
374. M. Broyer, G. Delacrétaz, N. Guoquan, J.P. Wolf, and L. Wöste, 'Lifetimes and Relaxation Processes in Electronically Exited States of Na_3', Chem. Phys. Lett. **145**, 232 (1988).
375. M. Broyer, G. Delacrétaz, P. Labastie, R.L. Whetten, and L. Wöste, 'Spectroscopy of Na_3', Z. Phys. D **3**, 131 (1986).
376. G. Delacrétaz, *Proprietés optiques d'agregats metalliques en phase gazeuse*, Ph.D. thesis, EPFL Lausanne, Lausanne, 1985.
377. W.E. Ernst and S. Rakowsky, 'Integer Quantization of the Pseudorotational Motion in Na_3 B', Phys. Rev. Lett. **74**, 58 (1995).
378. R. de Vivie-Riedle, J. Gaus, V. Bonačić-Koutecký, J. Manz, B. Reischl, S. Rutz, E. Schreiber, and L. Wöste, 'Pulse Width Controlled Molecular Dynamics: Symmetric Stretch Versus Pseudorotations in Na_3(B)' in *Femtochemistry: Ultrafast Chemical and Physical Processes in Molecular Systems*, M. Chergui (ed.) (World Scientific, Singapore, 1996), p 319.
379. K. Kobe, H. Kühling, S. Rutz, E. Schreiber, J.P. Wolf, L. Wöste, M. Broyer, and Ph. Dugourd, 'Time-Resolved Observation of Molecular Pseudorotation in Na_3', Chem. Phys. Lett. **213**, 554 (1993).
380. J. Gaus, K. Kobe, V. Bonačić-Koutecký, H. Kühling, J. Manz, B. Reischl, S. Rutz, E. Schreiber, and L. Wöste, 'Experimental and Theoretical Approach to the Pseudorotating Na_3 (B)', J. Phys. Chem. **97**, 12 509 (1993).

381. S. Rutz, H. Ruppe, E. Schreiber, and L. Wöste, 'Femtosecond Wave Packet Dynamics in Alkali Trimers', Z. Phys. D **40**, 25 (1997).
382. B. Reischl, R. de Vivie-Riedle, S. Rutz, and E. Schreiber, 'Ultrafast Molecular Dynamics Controlled by Pulse Duration: The Na_3 Molecule', J. Chem. Phys. **104**, 8857 (1996).
383. R.S. Berry, V. Bonačić-Koutecký, J. Gaus, T. Leisner, J. Manz, B. Reischl-Lenz, H. Ruppe, S. Rutz, E. Schreiber, Š. Vajda, R. de Vivie-Riedle, S. Wolf, and L. Wöste, 'Size Dependent Ultrafast Relaxation Phenomena in Metal Clusters' in *Advances in Chemical Physics – Chemical Reactions and their Control on the Femtosecond Time-Scale*, P. Gaspard, I. Burghardt, I. Prigogine, and S. A. Rice (eds.) (Wiley, New York, 1997), Vol. 101, p 101.
384. S. Carter and W. Meyer, 'A Variational Method for the Calculation of Vibrational Energy Levels of Triatomic Molecules Using a Hamiltonian in Hyperspherical Coordinates', J. Chem. Phys. **93**, 8902 (1990).
385. S. Carter and W. Meyer, 'A Variational Method for the Calculation of Rovibrational Energy Levels of Triatomic Molecules Using a Hamiltonian in Hyperspherical Coordinates: Applications H_3^+ and Na_3^+', J. Chem. Phys. **100**, 2104 (1994).
386. J. Schön and H. Köppel, 'Geometric Phase Effects and Wave Packet Dynamics on Intersecting Potential Energy Surfaces', J. Chem. Phys. **103**, 9292 (1995).
387. A.J. Dobbyn and J.M. Hutson, 'Wavepacket Calculations of Femtosecond Pump–Probe Experiments on the Sodium Trimer', J. Phys. Chem. **98**, 11 428 (1994).
388. B. Reischl and M. Miertschink, 'A Movie of the Time-Dependent 3d *Ab-Initio* Wave Packet for Na_3 (B)', see http://www.chemie.fu-berlin.de/fb-chemie/ipc/manzwww/www/default.html (1995).
389. J. Manz and C.S. Parmenter, 'Mode Selectivity in Unimolecular Reactions' in *Mode Selectivity in Unimolecular Reactions*, Vol. 139 *Special Issue, Chemical Physics*, J. Manz and C.S. Parmenter (eds.) (North-Holland, Amsterdam, 1989), pp. 1–239.
390. M. Broyer, G. Delacrétaz, P. Labastie, J.P. Wolf, and L. Wöste, 'Vibronic Structure of the Na_3 Ground State by Stimulated Emission Spectroscopy', Phys. Rev. Lett. **62**, 2100 (1989).
391. E. Schreiber and S. Rutz, in preparation.
392. E. Schreiber, K. Kobe, A. Ruff, S. Rutz, G. Sommerer, and L. Wöste, 'Ultrafast Fragmentation Probability of the Na_3 C State', Chem. Phys. Lett. **242**, 106 (1995).
393. M. Broyer, G. Delacrétaz, G.-Q. Ni, R.L. Whetten, J.P. Wolf, and L. Wöste, 'Stimulated Emission Spectroscopy of the Ground State of Na_3', J. Chem. Phys. **90**, 843 (1989).
394. T.W. Ducas, M.G. Liimann, M.L. Zimmermann, and D. Kleppner, 'Radiative Lifetimes of Selected Vibrational Levels in the $A^1\Sigma_u$ State of Na_2^*', J. Chem. Phys. **65**, 842 (1976).
395. W. Demtröder, W. Stetzenbach, M. Stock, and J. Witt, 'Lifetimes and Franck–Condon Factors for the $B^1\Pi_u \rightarrow X^1\Sigma_u^+$ System of Na_2', J. Mol. Spectrosc. **61**, 382 (1976).
396. W.D. Knight, K. Clemenger, W.A. de Heer, W.A. Saunders, M.Y. Chou, and M.L. Cohen, 'Electronic Shell Structure and Abundances of Sodium Clusters', Phys. Rev. Lett. **52**, 2141 (1984).
397. H. Kühling, S. Rutz, K. Kobe, E. Schreiber, and L. Wöste, 'Femtosecond Fragmentation of the Na_3 D-state', J. Phys. Chem. **97**, 12500 (1993).
398. S. Rutz, K. Kobe, H. Kühling, E. Schreiber, and L. Wöste, 'Time-Resolved TPI Spectroscopy of Na_3-Clusters', Z. Phys. D **26**, 276 (1993).

399. I. Boustani, W. Pewestorf, P. Fantucci, V. Bonačić-Koutecký, and J. Koutecký, 'Systematic *Ab Initio* Configuration Interaction Study of Alkali Metal Clusters: Relation between Electronic Structure and Geometry of Small Li Clusters', Phys. Rev. B **35**, 9437 (1987).
400. W.H. Andreoni, 'Computer Simulations of Small Semiconductor and Metal Clusters', Z. Phys. D **19**, 31 (1991).
401. W.A. de Heer, W. Knight, M.Y. Chou, and M.L. Cohen, 'Electronic Shell Structure and Metal Clusters', Solid State Phys. **40**, 93 (1987).
402. K. Selby, M. Vollmer, J. Masui, V. Kresin, W.A. de Heer, and W.D. Knight, 'Surface Plasma Resonances in Free Metal Clusters', Phys. Rev. B **40**, 5417 (1989).
403. T.P. Martin, T. Bergmann, H. Göhlich, and T. Lange, 'Electronic Shells and Shells of Atoms in Metallic Clusters', Z. Phys. D **19**, 25 (1991).
404. S. Bjørnholm, J. Borggreen, O. Echt, K. Hansen, and H.D. Rasmussen, 'The Influence of Shells, Electron Thermodynamics, and Evaporation on the Abundance Spectra of Large Sodium Metal Clusters', Z. Phys. D **19**, 31 (1991).
405. H. Kühling, S. Rutz, K. Kobe, E. Schreiber, and L. Wöste, 'Odd–Even Alternation of Femtosecond Fragmentation Processes of Excited Na_{3-10} Clusters', J. Phys. Chem. **98**, 6697 (1994).
406. C. Bréchignac, Ph. Cahuzac, F. Carlier, M. de Frutos, and J. Leygnier, 'Simple Metal Clusters', Z. Phys. D **19**, 1 (1991).
407. C. Wang, S. Pollak, and M.M. Kappes, 'Molecular Excited States versus Collective Electronic Oscillations: Optical Absorption Probes of Na_4 and Na_5', Chem. Phys. Lett. **166**, 26 (1990).
408. V. Bonačić-Koutecký, M. Kappes, P. Fantucci, and J. Koutecký, 'Interpretation of the Absorption Spectrum of Na_8', Chem. Phys. Lett. **170**, 26 (1990).
409. V. Bonačić-Koutecký, P. Fantucci, C.J. Koutecký, and J. Pittner, 'Nature of Exitations in Small Alkali Metal and Other Mixed Clusters', Z. Phys. D **26**, 17 (1993).
410. V. Bonačić-Koutecký, private communications.
411. A. Ruff, *Ultraschnelle Photodissoziation kleiner Alkalicluster*, Master's thesis, Freie Universität Berlin, Berlin-Dahlem, 1994.
412. F. Spiegelmann and D. Pavolini, 'Ab Initio Calculations of the Electronic Structure of Small Na_n, Na_n^+, K_n, and K_n^+ Clusters ($n \leq 6$) Including Core–Valence Interaction', J. Chem. Phys. **88**, 4954 (1988).
413. U. Buck, 'Structure and Dynamics of Small Size Selected Molecular Clusters', J. Phys. Chem. **98**, 5190 (1994).
414. R.B. Metz, S.E. Bradforth, and D.M. Neumark, 'Transition State Spectroscopy of Bimolecular Reactions Using Negative Ion Photodetachment', Adv. Chem. Phys. **81**, 1 (1992).
415. D.M. Neumark, 'Transition-State Spectroscopy via Negative Ion Photodetachment', Acct. Chem. Res. **26**, 33 (1993).
416. A. Weaver, R.B. Metz, S.E. Bradforth, and D.M. Neumark, 'Investigation of the $F+H_2$ Transition State Region via Photoelectron Spectroscopy of the FH_2^- Ion', J. Chem. Phys. **93**, 5352 (1990).
417. R.B. Metz and D.M. Neumark, 'Adiabatic Three-Dimensional Simulations of the IHI^-, $BrHI^-$, and $BrHBr^-$ Photoelectron Spectra', J. Chem. Phys. **97**, 962 (1992).
418. S. Wolf, G. Sommerer, S. Rutz, E. Schreiber, T. Leisner, and L. Wöste, 'Spectroscopy of Size-Selected Neutral Clusters: Femtosecond Evolution of Neutral Silver Trimers', Phys. Rev. Lett. **74**, 4177 (1995).

419. V. Bonačić-Koutecký, L. Češpiva, P. Fantucci, and J. Koutecký, 'Effective Core Potential-Configuration Interaction Study of Electronic Structure and Geometry of Small Neutral and Cationic Ag_n Clusters: Predictions and Interpretation of Measured Properties', J. Chem. Phys. **98**, 7981 (1993).
420. D.W. Boo, Y. Ozaki, L.H. Andersen, and W.C. Lineberger, 'Femtosecond Dynamics of Linear Ag_3', J. Phys. Chem. A **101**, 6688 (1997).
421. E. Schreiber, R.S. Berry, T. Leisner, S. Rutz, S. Wolf, and L. Wöste, 'Femtosecond Dynamics of the Ground State of Ag_3: A New Approach to Study the Ultrafast Dynamics of Mass-Selected Neutral Clusters' in *Ultrafast Processes in Spectroscopy*, O. Svelto, D. De Silvestri, and G. Denardo (eds.) (Plenum, New York, 1996), pp. 133–137.
422. M. Broyer, G. Delacrétaz, P. Labastie, J.P. Wolf, and L. Wöste, 'Spectroscopy of Vibrational Ground-State Levels of Na_3', J. Phys. Chem. **91**, 2626 (1987).
423. I.B. Bersuker, *The Jahn–Teller Effect and Vibronic Interactions in Modern Chemistry* (Plenum, New York, 1984).
424. T. Leisner, S. Rutz, S. Vajda, S. Wolf, E. Schreiber, and L. Wöste, 'Spectroscopy on Mass-Selected Neutral Clusters: Femtosecond Dynamics of the Ground State of Ag_n' in *Fast Elementary Processes in Chemical and Biological Systems*, Vol. 364 *AIP Conference Proceedings*, A. Tramer (ed.) (AIP Press, Woodbury, New York, 1996), p 603.
425. *Femtosecond Chemistry*, J. Manz and L. Wöste (eds.) (VCH, Weinheim, 1995), Vol. 1.
426. P. Gaspard and I. Burghardt (eds.), *Chemical Reaction and their Control on the Femtosecond Time Scale – XXth Solvay Conference on Chemistry*, Vol. 101 *Advances in Chemical Physics*, (Wiley, New York, 1997).
427. *Femtosecond Reaction Dynamics*, D.A. Wiersma (ed.) (North-Holland, Amsterdam, 1994).
428. A.H. Zewail, 'Femtochemistry: Recent Progress in Studies of Dynamics and Control of Reactions and Their Transition States', J. Phys. Chem. **100**, 12701 (1996).
429. G. Delacrétaz and L. Wöste, 'Two-Photon Ionization Spectroscopy of the $(2)^1\Sigma_u^1$ Double-Minimum State of Na_2', Chem. Phys. Lett. **120**, 342 (1985).
430. R. Haugstätter, A. Goerke, and I.V. Hertel, 'Case Studies in Multiphoton Ionisation and Dissociation of Na_2, I. The $(2)^1\Sigma_u$ Pathway', Z. Phys. D **9**, 153 (1988).
431. F. Stienkemeier, J. Higgins, W.E. Ernst, and G. Scoles, 'Laser Spectroscopy of Alkali-Doped Helium Clusters', Phys. Rev. Lett. **74**, 3592 (1995).
432. F. Stienkemeier, J. Higgins, C. Callegari, S.I. Kanorsky, W.E. Ernst, and G. Scoles, 'Spectroscopy of Alkali Atoms (Li, Na, K) Attached to Large Helium Clusters', Z. Phys. D **38**, 253 (1996).
433. J. Higgins, C. Callegari, J. Reho, F. Stienkemeier, W.E. Ernst, K.K. Lehmann, M. Gutowski, and G. Scoles, 'Photoinduced Chemical Dynamics of High-Spin Alkali Trimers', Science **273**, 629 (1996).
434. T. Wilhelm, J. Piel, and E. Riedle, 'Sub-20-fs Pulses Tunable Across the Visible from a Blue-Pumped Single-Pass Noncollinear Parametric Converter', Opt. Lett. **22**, 1494 (1997).
435. M.M. Wefers and K.A. Nelson, 'Space–Time Profiles of Shaped Ultrafast Optical Waveforms', IEEE Quant. Electr. **32**, 161 (1996).
436. G.J. Tóth, A. Lőrincz, and H. Rabitz, 'The effect of control field and measurement imprecision on laboratory feedback control of quantum systems', J. Chem. Phys. **101**, 3715 (1994).

437. P. Gross, D. Neuhauser, and H. Rabitz, 'Teaching lasers to control molecules in the presence of laboratory field uncertainity and measurement imprecision', J. Chem. Phys. **98**, 4557 (1993).
438. R.S. Judson and H. Rabitz, 'Teaching Lasers to Control Molecules', Phys. Rev. Lett. **68**, 1500 (1992).
439. J.A. China, 'Phase-Controlled Optical Pulses and the Adiabatic Electronic Sign Change', Phys. Rev. Lett. **66**, 1146 (1991).
440. W.S. Warren, 'Effects of Pulse Shaping in Laser Spectroscopy and Nuclear Magnetic Resonance', Science **242**, 878 (1988).
441. M.B. Danailov and I.P. Christow, 'Time–Space Shaping of Light Pulses by Fourier Optical Processing', J. Mod. Opt. **36**, 725 (1989).
442. A.M. Weiner, D.E. Leaird, J.S. Patel, and J.R. Wullert II, 'Programmable Shaping of Femtosecond Optical Pulses by Use of 128-Element Liquid Crystal Phase Modulator', IEEE Quant. Electr. **28**, 908 (1992).
443. C.W. Hillegas, J.X. Tull, D. Goswami, D. Strickland, and W.S. Warren, 'Femtosecond Laser Pulse Shaping by Use of Microsecond Radio-Frequency Pulses', Opt. Lett. **19**, 737 (1994).
444. Z. Wang, Z. Zhang, Z. Xu, and Q. Lin, 'Space–Time Profiles of an Ultrashort Pulsed Gaussian Beam', IEEE Quant. Electr. **33**, 566 (1997).
445. J. Manz, B. Proppe, and B. Schmidt, 'From Torsional Spectra to Hamiltonians and Dynamics: Effects of Coupled Bright and Dark States of 9-(N-Carbazolyl)-Anthracene', Z. Phys. D **34**, 111 (1995).
446. J. Giraud-Girard, J. Manz, and Ch. Scheurer, 'Twist Dynamics of 9-(N-Carbazolyl)-Anthracene: Effects of Intramolecular Vibrational Redistribution and Non-Adiabatic Transitions in Coupled Bright and Dark States', Z. Phys. D **39**, 291 (1997).
447. Ch. Monte, A. Roggon, A. Subaric-Leitis, W. Rettig, and P. Zimmermann, 'Resonance Effects of Diabatic Surface Crossing within the Torsional Spectrum of 9-(N-Carbazolyl) Anthracene Observed by Supersonic Jet Fluorescence Spectroscopy', J. Chem. Phys. **98**, 2580 (1993).
448. A. Subaric-Leitis, Ch. Monte, A. Roggan, W. Rettig, and P. Zimmermann, 'Torsional Band Assignment and Intramolecular Twist Potential of $9,9'$-Bianthryl and its 10-Cyano Derivate in a Free Jet', J. Chem. Phys. **93**, 4543 (1990).
449. K.M. Ervin, J. Ho, and W.C. Lineberger, 'A Study of the Singlet and Triplett States of Vinylidene by Photoelectron Spectroscopy of $H_2C{=}C^-$, $D_2C{=}C^-$, and $HDC{=}C^-$. Vinylidene-Acetylene Isomerization', J. Chem. Phys. **91**, 5974 (1989).

Index

Springer Tracts in Modern Physics

125 **Inelastic Scattering of X-Rays with Very High Energy Resolution**
By E. Burkel 1991. 70 figs. XV, 112 pages

126 **Critical Phenomena at Surfaces and Interfaces**
Evanescent X-Ray and Neutron Scattering
By H. Dosch 1992. 69 figs. X, 145 pages

127 **Critical Behavior at Surfaces and Interfaces**
Roughening and Wetting Phenomena
By R. Lipowsky, not yet published

128 **Surface Scattering Experiments with Conduction Electrons**
By D. Schumacher 1993. 55 figs. IX, 95 pages

129 **Dynamics of Topological Magnetic Solitons**
By V. G. Bar'yakhtar, M. V. Chetkin, B. A. Ivanov, and S. N. Gadetskii
1994. 78 figs. VIII, 179 pages

130 **Time-Resolved Light Scattering from Excitons**
By H. Stolz 1994. 87 figs. XI, 210 pages

131 **Ultrathin Metal Films**
Magnetic and Structural Properties
By M. Wuttig, not yet published

132 **Interaction of Hydrogen Isotopes with Transition-Metals and Intermetallic Compounds**
By B. M. Andreev, E. P. Magomedbekov, G. Sicking 1996. 72 figs. VIII, 168 pages

133 **Matter at High Densities in Astrophysics**
Compact Stars and the Equation of State
In Honor of Friedrich Hund's 100th Birthday
By H. Riffert, H. Müther, H. Herold, and H. Ruder 1996. 86 figs. XIV, 278 pages

134 **Fermi Surfaces of Low-Dimensional Organic Metals and Superconductors**
By J. Wosnitza 1996. 88 figs. VIII, 172 pages

135 **From Coherent Tunneling to Relaxation**
Dissipative Quantum Dynamics of Interacting Defects
By A. Würger 1996. 51 figs. VIII, 216 pages

136 **Optical Properties of Semiconductor Quantum Dots**
By U. Woggon 1997. 126 figs. VIII, 252 pages

137 **The Mott Metal-Insulator Transition**
Models and Methods
By F. Gebhard 1997. 38 figs. XVI, 322 pages

138 **The Partonic Structure of the Photon**
Photoproduction at the Lepton-Proton Collider HERA
By M. Erdmann 1997. 54 figs. X, 118 pages

139 **Aharonov-Bohm and other Cyclic Phenomena**
By J. Hamilton 1997. 34 figs. X, 186 pages

140 **Exclusive Production of Neutral Vector Mesons at the Electron-Proton Collider HERA**
By J. A. Crittenden 1997. 34 figs. VIII, 108 pages

141 **Disordered Alloys**
Diffuse Scattering and Monte Carlo Simulations
By W. Schweika 1998. 48 figs. X, 126 pages

142 **Phonon Raman Scattering in Semiconductors, Quantum Wells and Superlattices**
Basic Results and Applications
By T. Ruf 1998. 143 figs. VIII, 252 pages

143 **Femtosecond Real-Time Spectroscopy of Small Molecules and Clusters**
By. E. Schreiber 1998. 131 figs. X, 212 pages

144 **New Aspects of Electromagnetic and Acoustic Wave Diffusion**
By POAN Research Group 1998. 30 figs. IX, 117 pages

MIX
Papier aus verantwortungsvollen Quellen
Paper from responsible sources
FSC® C105338

If you have any concerns about our products,
you can contact us on
ProductSafety@springernature.com

In case Publisher is established outside the EU,
the EU authorized representative is:
Springer Nature Customer Service Center GmbH
Europaplatz 3, 69115 Heidelberg, Germany

Printed by Libri Plureos GmbH
in Hamburg, Germany